The Structure and Properties of Color Spaces and the Representation of Color Images

The Structure and Properties of Color Spaces and the Representation of Color Images
Eric Dubois

ISBN: 978-3-031-01118-4   paperback
ISBN: 978-3-031-02246-3   ebook

DOI 10.1007/978-3-031-02246-3

A Publication in the Springer series
SYNTHESIS LECTURES ON IMAGE, VIDEO, AND MULTIMEDIA PROCESSING

Lecture #11
Series Editor: Alan C. Bovik, *University of Texas, Austin*
Series ISSN
Synthesis Lectures on Image, Video, and Multimedia Processing
Print 1559-8136   Electronic 1559-8144

Synthesis Lectures on Image, Video, and Multimedia Processing

Editor
Alan C. Bovik, *University of Texas, Austin*

The Structure and Properties of Color Spaces and the Representation of Color Images
Eric Dubois
2009

Biomedical Image Analysis: Segmentation
Scott T. Acton, Nilanjan Ray
2009

Joint Source-Channel Video Transmission
Fan Zhai, Aggelos Katsaggelos
2007

Super Resolution of Images and Video
Aggelos K. Katsaggelos, Rafael Molina, Javier Mateos
2007

Tensor Voting: A Perceptual Organization Approach to Computer Vision and Machine Learning
Philippos Mordohai, Gérard Medioni
2006

Light Field Sampling
Cha Zhang, Tsuhan Chen
2006

Real-Time Image and Video Processing: From Research to Reality
Nasser Kehtarnavaz, Mark Gamadia
2006

The Structure and Properties of Color Spaces and the Representation of Color Images

Eric Dubois
University of Ottawa

SYNTHESIS LECTURES ON IMAGE, VIDEO, AND MULTIMEDIA PROCESSING #11

ABSTRACT

This lecture describes the author's approach to the representation of color spaces and their use for color image processing. The lecture starts with a precise formulation of the space of physical stimuli (light). The model includes both continuous spectra and monochromatic spectra in the form of Dirac deltas. The spectral densities are considered to be functions of a continuous wavelength variable.

This leads into the formulation of color space as a three-dimensional vector space, with all the associated structure. The approach is to start with the axioms of color matching for normal human viewers, often called Grassmann's laws, and developing the resulting vector space formulation. However, once the essential defining element of this vector space is identified, it can be extended to other color spaces, perhaps for different creatures and devices, and dimensions other than three. The CIE spaces are presented as main examples of color spaces. Many properties of the color space are examined.

Once the vector space formulation is established, various useful decompositions of the space can be established. The first such decomposition is based on luminance, a measure of the relative brightness of a color. This leads to a direct-sum decomposition of color space where a two-dimensional subspace identifies the chromatic attribute, and a third coordinate provides the luminance. A different decomposition involving a projective space of chromaticity classes is then presented. Finally, it is shown how the three types of color deficiencies present in some groups of humans leads to a direct-sum decomposition of three one-dimensional subspaces that are associated with the three types of cone photoreceptors in the human retina. Next, a few specific linear and nonlinear color representations are presented. The color spaces of two digital cameras are also described. Then the issue of transformations between *different* color spaces is addressed.

Finally, these ideas are applied to signal and system theory for color images. This is done using a vector signal approach where a general linear system is represented by a three-by-three system matrix. The formulation is applied to both continuous and discrete space images, and specific problems in color filter array sampling and displays are presented for illustration.

The book is mainly targeted to researchers and graduate students in fields of signal processing related to any aspect of color imaging.

KEYWORDS

color, color space, color image, colorimetry, color Fourier transform, color filter array, color mosaic display, color image filtering, color image sampling

To my colorful grandchildren

Kiernan, Quinn, and Juliette

Contents

Preface

Color is everywhere and electronic color imaging is now pervasive. Color is an aspect of human perception and the attempt to describe it mathematically has been going on for centuries, with important contributions from great scientists like Newton, Maxwell, Helmholz, Schrödinger and many, many more [32]. There are numerous excellent, comprehensive and beautiful treatises on color theory, color science, colorimetry and so on ([61], [21], [28], [65] to name just a few). However, it is my belief that none of these adequately present the algebraic theory of color spaces and their applications in color imaging in a suitably comprehensive fashion for signal processing theorists. This lecture describes my approach to the representation of color spaces and their use for color image processing. It is mainly targeted to researchers and graduate students in fields of signal processing related to any aspect of color imaging. However, I will be happy if it finds use in other aspects of color science in general. I have attempted to establish a complete and consistent notation to deal with all aspects of this subject. Since there is no single universally adopted notation for many of the concepts used here, I had to introduce new notation for many aspects. However, I have tried to remain consistent with all standard nomenclature such as color matching functions, chromaticity, etc. All the notation has been summarized in the Notation section, along with a synthetic diagram illustrating the main sets, spaces and transformations that have been introduced.

I would like to thank several colleagues who have read various versions of this book and provided valuable corrections and suggestions for improvement. Specifically, my great thanks to Prof. Stéphane Coulombe of the Department of Software and IT Engineering, École de technologie supérieure, Montreal, Quebec; Prof. Abdol-Reza Mansouri of the Department of Mathematics and Statistics, Queen's University, Kingston Ontario; and Prof. Gaurav Sharma of the Department of Electrical and Computer Engineering, University of Rochester, Rochester, New York. Their input pushed me to make many improvements to the book, as well to correct a number of errors, and I have attempted to address all the issues they raised. Of course, all the remaining errors and inadequacies are entirely my responsibility. I also thank Mustafa Fanaswala for pointing out a number of errors.

In addition, I would like thank several people who provided data or clarifications. I thank Cong Phuoc Huynh, of the Department of Information Engineering, College of Engineering and Computer Science, Australian National University, Canberra, Australia, for his data on the spectral sensitivity of digital cameras; Charles Poynton, noted expert on color imaging systems of Toronto, Ontario, for data on the spectral density of red, green and blue phosphors in a CRT display; and Xuemei Zhang, senior researcher at HP labs in Palo Alto, California, for explanations concerning computation of the S-CIELAB error measure. Finally, I thank Joel Claypool and Al Bovik for their encouragement to complete this work and their patience in waiting for the final result while I was busy with my administrative duties.

Eric Dubois
October 2009

Notation

Radiometry and Photometry

$P_e, P_e(\lambda)$	Radiant power (W), radiant power spectral density (W/nm)
$E_e, E_e(\lambda)$	Irradiance (W/m^2), (W/m^{2}nm)
$M_e, M_e(\lambda)$	Radiant exitance (W/m^2), (W/m^{2}nm)
$I_e, I_e(\lambda)$	Radiant intensity (W/sr), (W/sr nm)
$L_e, L_e(\lambda)$	Radiance (W/sr m^2), (W/sr m^2 nm)
L_v	Luminance (lm/sr m^2)

Equivalence Relations

$\triangleq$	Trichromatic physical metameric equivalence on $\mathcal{P}$
$\boxminus$	Trichromatic extended metameric equivalence on $\mathcal{A}$
$\triangle$	Brightness equivalence on $\mathcal{P}$
$\boxdot$	Extended brightness equivalence on $\mathcal{A}$
$\boxasymp$	Dichromatic extended metameric equivalence on $\mathcal{A}$

Sets and Spaces

$\mathcal{P}$	The set of physical light stimuli
$\mathcal{A}$	The vector space of differences of physical light stimuli
$\mathcal{A}_c, \mathcal{A}_d$	The continuous and discrete subspaces of $\mathcal{A}$ in the decomposition $\mathcal{A} = \mathcal{A}_c \oplus \mathcal{A}_d$
$\mathcal{A}^*$	Dual space of $\mathcal{A}$, space of linear functionals from $\mathcal{A}$ to $\mathbb{R}$
$\mathcal{A}^*_{\mathscr{C}}$	Subspace of $\mathcal{A}^*$ consisting of elements continuous on $\mathcal{V}$
$\mathcal{V}$	The set of wavelengths comprising the visible spectrum, e.g., [350 nm, 780 nm]
$\mathcal{C}$	Color vector space
$\mathcal{D}_P, \mathcal{D}_D, \mathcal{D}_T$	Generic dichromat color spaces for protanopes, deuteranopes and tritanopes, respectively
$[C(\lambda)]_\sim$	Equivalence class of spectral densities under the equivalence relation $\sim$
$\mathcal{B}$	A generic basis for the color space $\mathcal{C}$
$\mathcal{K}$	Black space, kernel (or null space) of the mapping $\mathcal{S}$
$\mathcal{VS}_{\mathcal{C}}$	Subspace of $\mathcal{A}^*_{\mathscr{C}}$ that defines the color space $\mathcal{C}$, spanned by any set of color matching functions
$\mathcal{C}_R$	The subset of $\mathcal{C}$ of physically realizable colors, $\mathcal{C}_R = \mathcal{S}(\mathcal{P})$
$\mathcal{M}$	The subset of $\mathcal{C}$ consisting of monochromatic lights
$\mathcal{H}$	The convex hull of $\mathcal{M}$
$\mathcal{F}([\mathbf{Q}])$	Physical metamer set, the subset of $\mathcal{P}$ consisting of all the physically realizable metamers of $[\mathbf{Q}]$ in a given color space

$\mathscr{M}(\cdot)$	Metamer mismatch set

Elements of Sets and Spaces

$[\mathbf{C}]$	Element of a color space, called simply a color
$\mathbf{C}_{\mathcal{B}}$	Column matrix of coefficients of color $[\mathbf{C}]$ with respect to basis $\mathcal{B}$ (tristimulus values)
$\boldsymbol{x}$	Independent variables, $(x, y) \in \mathbb{R}^2$ for still images, or $(x, y, t) \in \mathbb{R}^2 \times \mathbb{R}$ for time-varying images
$\boldsymbol{u}$	Frequency variables, (u, v) for still images, or (u, v, w) for time-varying images
$[\mathbf{C}](\boldsymbol{x})$	Element of a color signal space, or value of a color signal at $\boldsymbol{x}$ (according to context)
$\mathbf{C}_{\mathcal{B}}(\boldsymbol{x})$	Color signal expressed with respect to basis $\mathcal{B}$
$[\widehat{\mathbf{C}}](\boldsymbol{u})$	Fourier transform of $[\mathbf{C}](\boldsymbol{x})$
$\widehat{\mathbf{C}}_{\mathcal{B}}(\boldsymbol{u})$	Fourier transform of $\mathbf{C}_{\mathcal{B}}(\boldsymbol{x})$ expressed in basis $\mathcal{B}$

Transformations

$\mathcal{S}$	Mapping from the vector space $\mathcal{A}$ of radiometric functions to the color space $\mathcal{C}$
$\mathbf{A}_{\mathcal{B}\to\tilde{\mathcal{B}}}$	Matrix to convert tristimulus values from basis $\mathcal{B}$ to basis $\tilde{\mathcal{B}}$
$\mathcal{T}_d$	Shift system
$\mathbf{H}_{\mathcal{B}}(\boldsymbol{u})$	Transfer matrix of a linear shift-invariant system expressed in basis $\mathcal{B}$

Special Functions

$\{\bar{p}_i(\lambda)\}$	Set of color matching functions relative to primaries $\{[\mathbf{P}_i]\}$ that specify the mapping $\mathcal{S}$
$V(\lambda)$	CIE photopic spectral luminous efficiency curve

Specific Color Spaces and Bases

$\mathcal{XYZ}$	Standard basis for the CIE 1931 colorimetric observer
$\mathcal{RGB}31$	Standard monochromatic RGB basis for the CIE 1931 colorimetric observer
$\mathcal{XYZ}_{10}$	Standard basis for the CIE 1964 10° colorimetric observer
$\mathcal{UCS}76$	CIE 1976 Uniform Chromaticity Scale (UCS) basis
$\mathcal{R}709$	RGB basis for CIE 1931 colorimetric observer standardized by ITU-R Rec. BT.709 for HDTV and also sRGB for multimedia
$\mathcal{OZW}$	Opponent-colors basis for the CIE 1931 colorimetric observer used on the S-CIELAB error metric
$\mathcal{C}_{\mathrm{SB}10}$	The Stiles and Burch 10° colorimetric observer
$\mathcal{SB}10$	The RGB primaries for the Stiles and Burch 10° colorimetric observer
$\mathcal{SS}10$	The Stockman and Sharp LMS primaries for the Stiles and Burch 10° colorimetric observer

Acronyms

CIE	Commission International d'Éclairage, or in English, International Commission on Illumination
sRGB	Default nonlinear RGB representation for multimedia, defined in IEC 61966-2-1
IEC	International Electrotechnical Commission
ITU	International Telecommunication Union
NTSC	National Television System Committee

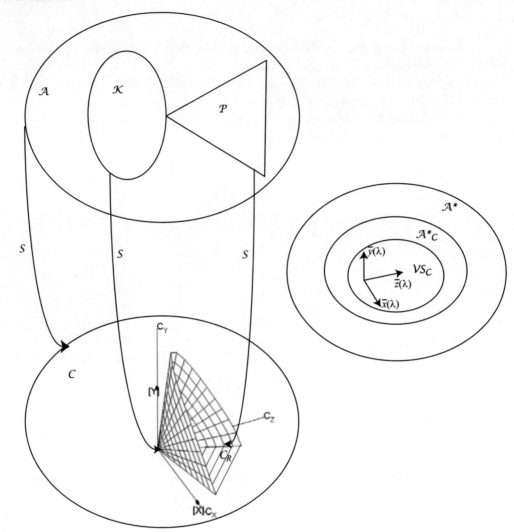

$\mathcal{P}$ is a subset of $\mathcal{A}$ (not a subspace), which is a convex cone

$\mathcal{K}$ is a vector subspace of $\mathcal{A}$

$\mathcal{K} \cap \mathcal{P} = 0$

$\mathcal{S}(\mathcal{A}) = \mathcal{C}, \mathcal{S}(\mathcal{P}) = \mathcal{C}_R, \mathcal{S}(\mathcal{K}) = 0$

$\mathcal{A}$ and $\mathcal{K}$ are infinite dimensional; $\dim(\mathcal{C}) = 3$

$\mathcal{A}^*$ is the dual space of $\mathcal{A}$; $\mathcal{A}^*_\mathcal{C}$ is a vector subspace of $\mathcal{A}^*$; $\mathcal{VS}_\mathcal{C}$ is a vector subspace of $\mathcal{A}^*_\mathcal{C}$

$\mathcal{A}^*$ and $\mathcal{A}^*_\mathcal{C}$ are infinite dimensional; $\dim(\mathcal{VS}_\mathcal{C}) = 3$

CHAPTER 1

Introduction

I have been working in color image and video processing since 1977. The initial work was with NTSC video signals. From the beginning, I have been curious about color spaces and color image spaces and had many questions, especially from a signal processing perspective. For example, it is well known that colors are a property of human vision and that a color can be considered to be an equivalence class of all radiometric spectral densities that appear identical to a human viewer. Since I did not find this representation rigorously presented in any of the literature I consulted, I started developing it for my image and video processing course. That development forms the nucleus of this lecture. It turns out that D.H. Krantz had, in fact, presented many of these ideas in 1975 [26] and so my work has much in common with his. However, this lecture presents much related work not covered by Krantz, and the approach is quite different. In particular, I did not use the concept of Grassmann structures in this lecture. All proofs presented in this lecture are my own. The section on color image spaces has no correspondence in Krantz's work.

I have attempted to present a complete picture of the underpinnings of the theory of color spaces and color image spaces, starting from a precise formulation of the space of physical stimuli (light) in Chapter 2. Here, the set of physically realizable radiometric functions is embedded in a vector space of differences of such physical signals. The model includes both continuous spectra and monochromatic spectra in the form of Dirac deltas. The spectral densities are considered to be functions of a continuous wavelength variable, and discretization is only introduced for the purpose of computation. This is different from most works that assume from the beginning a finite-dimensional model for the space of radiometric function [53], [46], [3].

This leads into the formulation in Chapter 3 of color space as a three-dimensional vector space, with all the associated structure. This does not seem to appear elsewhere in such detail, outside the work of Krantz. I have attempted to answer most of the questions that I have posed myself over the years, generally in the form of theorems with proof. All results directly pertaining to color spaces and color image representation are designated as theorems, independently of the relative importance of the particular result. The approach has been to start with the axioms of color matching for normal human viewers, often called Grassmann's laws, and developing the resulting vector space formulation. However, once the essential defining element of this vector space is identified, it can be extended to other color spaces, perhaps for different creatures and devices, and dimensions other than three. I try to make clear the distinction between *color space*, which is associated with a given individual, group of individuals, device or model, and a *coordinate representation* for a given color space. It is common to refer to such coordinate representations as color spaces. The CIE spaces are presented as main examples of color spaces. It is also important to distinguish between the *representation* of a color, which is done using a set of (usually three) real numbers, and the physical *synthesis* of a color, which is a process.

Once the vector space formulation is established, various useful decompositions of the space can be established. Several of these are presented in Chapter 4. The first such decomposition is based on luminance, a measure of the relative brightness of a color. This leads to a direct-sum decomposition of color space, where a two-dimensional subspace identifies the chromatic attribute, and a third coordinate provides the luminance. A different decomposition involving a projective space of chromaticity classes is then presented. Finally, it is shown how the three types of color deficiencies present in some groups of humans leads to a direct-sum decomposition of three one-dimensional subspaces that are associated with the three types of cone photoreceptors in the human retina.

In Chapter 5, a few specific linear and nonlinear color representations are presented. The color spaces of two digital cameras are also described. Then the issue of transformations between *different* color spaces is addressed. Such a transformation cannot, in general, be done exactly since the different color spaces do not consist of the same equivalence classes. Thus, an error measure must be minimized. The structure of the transformation is presented, but specific optimizations are not considered here.

Finally, in Chapter 6, these ideas are applied to signal and system theory for color images. In practice, a system for color images is usually implemented using separate scalar systems on three coordinate images. Here, I present the vector signal approach, where a general linear system is represented by a three-by-three system matrix. Only for a particular choice of basis may this matrix be diagonal as assumed in the scalar approach. The formulation is applied to both continuous and discrete space images, and specific problems in color filter array sampling and displays are presented for illustration.

It is my hope that this lecture will be useful to researchers in signal processing and will provide a solid foundation for color image processing. Many topics in this lecture can be developed much further but due to time and length constraints, I have limited it to the material presented herein. It is important to emphasize that this lecture does not contain any new contributions to the modeling of human vision, but it is hoped that the content can be adapted as these models are further developed.

The main axioms of color matching are referred to as Grassmann's laws, referring to his classic paper of 1853 [16]. It is most fitting that Grassmann is also credited with the invention of linear algebra [14], and that 2009 is the 200th anniversary of his birth.

CHAPTER 2

Light: The Physical Color Stimulus

2.1 BASIC RADIOMETRIC CONCEPTS

Light is the physical stimulus that gives rise to color perception when it arrives on the retinas of the human eyes. For our purposes, light is electromagnetic radiation with wavelengths roughly in the band from 350 to 780 nm. This is the range of wavelengths to which the human eye is sensitive and has a response (although response below 400 nm and above 700 nm is very small). Other branches of photonics give a broader interpretation of the term 'light,' and there is no precise definition. Light propagates in the form of electromagnetic waves that can be described in detail in terms of the variation of the electric and magnetic fields in space and time. However, most image sensors including the human eye respond to the energy or power of this radiation, and so it is sufficient to characterize this energy or power.

The measurement of light radiant energy, independent of its effect on the human visual system, is known as *radiometry*. This radiant energy is denoted Q_e, measured in Joules (J). In color perception, we are generally more concerned with the *power* of light passing through a given surface (say the retina, an electronic image sensor, or a display screen) at a given place and time than in total energy. There are a number of standard units that measure various forms of radiant power. *Radiant power* (also called radiant flux), denoted P_e, is the radiant energy emitted, transferred or received by a given surface per unit of time. We can write $P_e = \frac{dQ_e}{dt}$, measured in Watts (W).

The electromagnetic waves that form the light ray can be described through Fourier analysis in terms of the constituent spectral components. The spectral distribution of light as a function of optical wavelength is described by the *power density spectrum* $P_e(\lambda)$, where $P_e(\lambda)\,d\lambda$ is the incremental power contained in the range of wavelengths $[\lambda, \lambda + d\lambda)$; $P_e(\lambda)$ is generally expressed in units of W/nm. Of course power is non-negative, so $P_e(\lambda) \geq 0$. Total power is given by $P_e = \int_0^\infty P_e(\lambda)\,d\lambda$, but we often write $P_e = \int_{\lambda_{\min}}^{\lambda_{\max}} P_e(\lambda)\,d\lambda$ to limit our consideration to wavelengths in a given range of interest, e.g., 350 to 780 nm.

Radiant flux refers to an entire surface, say a whole image sensor or CRT display. We are usually interested in more localized measures. For example *irradiance* E_e measures the radiant flux per unit area falling on a surface at a particular location, $E_e = \frac{dP_e}{dA}$, measured in W/m². This would be relevant in measuring the relative intensity of light falling at a particular point on an image sensor. If the surface is planar and has an associated spatial coordinate system, we denote the irradiance at position (x, y) at time t by $E_e(x, y, t)$. A closely related concept is *radiant exitance*, denoted $M_e = \frac{dP_e}{dA}$ in W/m², measuring the radiant flux per unit area emitted by a surface at a particular location. This would be relevant in measuring the relative intensity of light emitted by a CRT display at a particular point on the screen.

Two final quantities of interest are *radiant intensity* and *radiance*. Radiant intensity, denoted I_e, is the radiant flux emitted by a point source of light in a given direction, per unit of solid angle, $I_e = \frac{dP_e}{d\omega}$, measured in W/sr, where sr stands for steradian, the unit of solid angle. A solid angle of 1 sr subtends a surface area of unity on a unit sphere. Radiance, L_e, is the radiant flux emitted by a surface in a given direction per unit of area and solid angle, $L_e = \frac{dP_e^2}{dA \cos \epsilon \, d\omega}$. All of these radiometric quantities can be represented as power density spectra, as a function of λ, so that the complete description of irradiance, for example, would be $E_e(x, y, t, \lambda)$. A good discussion of these radiometric concepts with nice illustrations can be found in [35].

The light arriving at a sensor can arise in a number of different ways. It can come directly from an emissive source like the sun, a flame or a light bulb. The radiation from such a source can pass through an optical filter like the red, green and blue filters in an LCD display or a piece of stained glass. Alternatively, the light can be reflected from a surface like a photograph or the objects around us. The physical mechanism underlying the origin of a light is very important in many applications including computer vision and computer graphics, and separation of the original light source from transmissive and reflective components can be a key task. However, this aspect is not central to the specification of color spaces and so will not be emphasized much in this lecture. However, everything in this lecture is compatible with such decompositions of the light.

Fig. 2.1 illustrates the power density spectra of several light sources of interest, including a halogen light source, a red helium-neon laser, and the red, green and blue lights emitted by typical CRT and LCD displays. In the case of the laser, at the scale of these figures, there is only a single wavelength present, 633 nm in this case. We refer to this as a monochromatic light. It can be approximated by a Dirac delta, as discussed in the next section. However, in reality, there is some spread. A typical value of the spread for this laser would be 10^{-8} nm, showing that on the scale of hundreds of nanometers, the Dirac delta is a good representation of the laser spectrum. Considerably more detail on radiometric concepts and data can be found in [61] and [1]. The discussion above is sufficient for our purposes.

2.2 THE SPACE OF PHYSICAL STIMULI

2.2.1 THE SET $\mathcal{P}$ OF PHYSICAL LIGHT STIMULI

In color perception, the physical stimulus is a ray of incoherent light arriving at a particular location on the sensor at a particular time. For this purpose, the light ray is completely characterized by its power density spectrum, and so the set of physical stimuli can be put in a one-to-one correspondence with the set of admissible power density spectra, and we consider the two sets to be equivalent. Let $f(x, y, t, \lambda)$ denote the power density spectrum of a physical light stimulus incident on the retina as a function of space and time, where x and y are in appropriate spatial units such as visual angle in degrees. This function describes the spectral composition of a light ray arriving at the coordinate (x, y) at time t as a function of wavelength λ and is a sufficient characterization of the light to

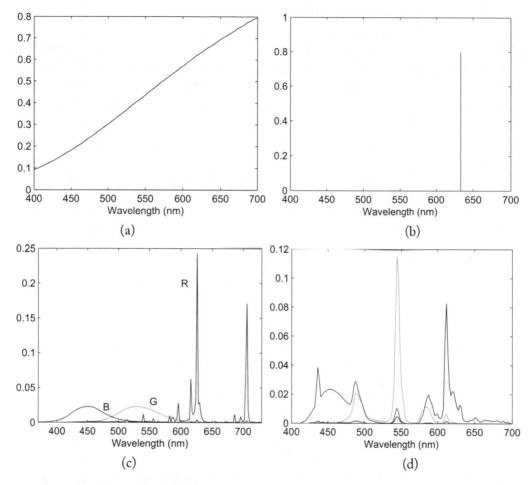

Figure 2.1: Illustration of power density spectra in arbitrary units for several light sources of interest. (a) Halogen light source. (b) Helium-neon laser at 633 nm. (c) Red, green and blue phosphors of a Sony trinitron CRT display. (d) Red, green and blue outputs of a SyncMaster LCD display.

determine the response of the visual system. We can assume that f represents irradiance and that it is real and non-negative. In the basic theory of color spaces, f is usually a fixed, constant patch of a given light, so that in the region of interest we can assume *for now* that there is no x, y or t variation, and simply write $f(\lambda)$. Since we are only concerned with wavelengths in the range $\mathcal{V} = [\lambda_{min}, \lambda_{max}] = \{\lambda | \lambda_{min} \leq \lambda \leq \lambda_{max}\}$, we could define a space of physical stimuli as the set of all non-negative, piecewise-continuous functions on $\mathcal{V}$.

Although we can reasonably expect that all physical light stimuli are non-negative, bounded and continuous on $\mathcal{V}$, it is convenient to assume that our signal space also contains the Dirac delta functions, as these are a good model for monochromatic lights, and many conventional light sources such as fluorescent lights exhibit spectral components at discrete wavelengths [47] (see Fig. 2.1(b) and Fig. 2.2(a)). We also can allow power density spectra with a finite number of discontinuities (or in other words, continuous for almost all λ) to include the possibility of lights being passed through ideal band-pass filters. Following the standard approach used for probability distributions and power density spectra of random processes [41], we assume that any physical light power density spectrum of interest can be written as the sum of two terms, one whose integrated spectrum is non-decreasing, continuous and bounded, and a second whose integrated spectrum is a non-decreasing step function with a finite number of steps. Let $F(\lambda)$ denote the integrated spectrum, $F(\lambda) = \int_{\lambda_{\min}}^{\lambda} f(\mu) \, d\mu$. The fact that $f(\lambda)$ is non-negative implies that $F(\lambda)$ is non-decreasing. We choose as signal space the set of all functions f such that $F(\lambda)$ can be written (in a unique fashion) as $F(\lambda) = F_c(\lambda) + F_d(\lambda)$, where $F_c(\lambda)$ is *absolutely continuous* (continuous everywhere and differentiable for almost all λ), and $F_d(\lambda)$ is a *step function* with steps at a finite set of wavelengths $\lambda_1, \lambda_2, \ldots, \lambda_K$, as illustrated in Fig. 2.2 for a typical warm white fluorescent lamp. See Section D.13 of [33] for some properties of absolutely continuous functions. Note that $F(\lambda_{\max})$ is the total power of $f(\lambda)$ in $\mathcal{V}$, assumed to be finite. It follows that the light power density spectrum can be written

$$f(\lambda) = f_c(\lambda) + \sum_{k=1}^{K} \alpha_k \delta(\lambda - \lambda_k), \quad f_c(\lambda) \geq 0, \lambda_k \in \mathcal{V}, \alpha_k \in \mathbb{R}_+^*, \tag{2.1}$$

where $\mathbb{R}_+^*$ is the set of strictly positive real numbers; $K = 0$ signifies that the second term is not present. The Dirac delta δ_{λ_k}, which is commonly written in functional form as $\delta_{\lambda_k}(\lambda) = \delta(\lambda - \lambda_k)$, is a generalized function that models very narrowband (i.e., monochromatic) lights. It is characterized by the property that

$$\int_{\lambda_{\min}}^{\lambda_{\max}} g(\mu)\delta(\mu - \lambda_0) \, d\mu = g(\lambda_0), \quad \lambda_0 \in \mathcal{V}, \tag{2.2}$$

for any function g that is continuous at $\lambda = \lambda_0$. The Dirac delta δ_{λ_k} can be considered as the limit as $\Delta \to 0$ of a rectangular pulse of width Δ and height $1/\Delta$ centered at λ_k. See Chapter 2 of [1] for a detailed description of the Dirac delta and its properties from the point of view of imaging systems. The Dirac delta is properly studied in the framework of the theory of distributions; see [42] for a rigorous but accessible introduction. In summary, we define the space $\mathcal{P}$ of physical light rays to be

$$\mathcal{P} = \left\{ f_c + \sum_{k=1}^{K} \alpha_k \delta_{\lambda_k} \ \middle| \ \begin{array}{l} \int_{\lambda_{\min}}^{\lambda} f_c(\mu) \, d\mu \text{ is non-decreasing and abso-} \\ \text{lutely continuous, } \lambda, \lambda_k \in \mathcal{V}, \alpha_k \in \mathbb{R}_+^*, K \in \mathbb{N} \end{array} \right\}, \tag{2.3}$$

where $\mathbb{N}$ denotes the set of natural numbers.

2.2.2   ALGEBRAIC STRUCTURE OF THE SET $\mathcal{P}$

We have identified a set of power density spectra that are in a one-to-one correspondence with physical light rays. We assume that these light rays consist of incoherent radiation. This implies that

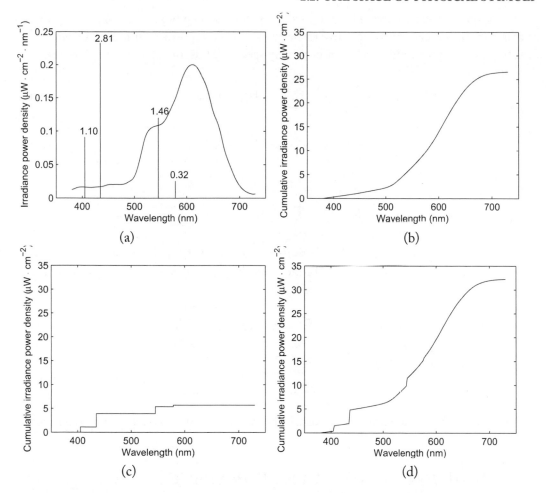

Figure 2.2: Illustration of the power density spectrum for a fluorescent light with both continuous and discrete components. (a) Irradiance power density spectrum. The irradiance is indicated for the monochromatic components. (b) Continuous component of the cumulative irradiance power density. (c) Step component of the cumulative irradiance power density. (d) Total cumulative irradiance power density corresponding to the irradiance in (a).

if two lights with power spectral densities $f_1(\lambda)$ and $f_2(\lambda)$ are superimposed, the power spectral density of the resulting light ray is $f_1(\lambda) + f_2(\lambda)$. Given this fact, we can start to impose some algebraic structure on the set $\mathcal{P}$. With this addition operation, we can state the following:

Theorem 2.1 $(\mathcal{P}, +)$ *is a commutative semigroup with a neutral element (or Abelian monoid) in which every element is cancellable.*

Proof. A semigroup is a set equipped with a binary, associative operation. See Appendix A for a formal definition. In our case, the associative operation is the superposition of light rays, which is equivalent to point-by-point addition of power density spectra. Thus, if

$$f_1(\lambda) = f_{1c}(\lambda) + \sum_{k=1}^{K_1} \alpha_{1k}\delta(\lambda - \lambda_{1k}) \tag{2.4}$$

$$f_2(\lambda) = f_{2c}(\lambda) + \sum_{k=1}^{K_2} \alpha_{2k}\delta(\lambda - \lambda_{2k}), \tag{2.5}$$

then

$$(f_1 + f_2)(\lambda) = (f_{1c}(\lambda) + f_{2c}(\lambda)) + \sum_{k=1}^{K_1} \alpha_{1k}\delta(\lambda - \lambda_{1k}) + \sum_{k=1}^{K_2} \alpha_{2k}\delta(\lambda - \lambda_{2k}), \tag{2.6}$$

which is of the required form. It is straightforward to show that $f_{1c} + f_{2c}$ is non-negative and continuous for almost all λ if f_{1c} and f_{2c} are. For the Dirac deltas, if $\lambda_{1i} = \lambda_{2j}$ for any pair (i, j), the two terms are combined into a single one with coefficient $\alpha_{1i} + \alpha_{2j}$. Thus we conclude that $f_1 + f_2 \in \mathcal{P}$ so that $\mathcal{P}$ is closed under the binary operation. The operation is also associative and commutative from the ordinary properties of the real numbers and distributions. The function $f(\lambda) = 0$ serves as the neutral element. Without ambiguity, we can write $(f_1 + f_2)(\lambda) = f_1(\lambda) + f_2(\lambda)$.

An element f is *cancellable* if for all $f_1, f_2 \in \mathcal{P}$

$$f + f_1 = f + f_2 \quad \Rightarrow \quad f_1 = f_2. \tag{2.7}$$

This holds for all $f \in \mathcal{P}$ in a straightforward fashion from the properties of real numbers and Dirac deltas. □

2.3   EMBEDDING OF $\mathcal{P}$ IN A VECTOR SPACE $\mathcal{A}$

Because of the positivity constraint on elements of $\mathcal{P}$, it is not possible in general to express the difference between two spectra as elements of this space. Thus, it is useful to extend $\mathcal{P}$ by embedding it in a larger algebraic structure that contains it, and whose operations are consistent with it. In this case, we will show how $\mathcal{P}$ can be embedded in a vector space $\mathcal{A}$, which will greatly assist in the representation of lights and colors. Informally, $\mathcal{A}$ consists of all the sums and differences of elements of $\mathcal{P}$.

First, we show that the semigroup $\mathcal{P}$ can be embedded in a unique way in a minimal group $\mathcal{A}$ such that every element of $\mathcal{P}$ has an additive inverse in $\mathcal{A}$. A group is a semigroup with a neutral element such that every element is invertible (see Appendix A). The group $\mathcal{A}$ is called the inverse completion of $(\mathcal{P}, +)$.

Theorem 2.2 *The semigroup $\mathcal{P}$ of physical light rays can be embedded in a unique, minimal commutative group $\mathcal{A}$ such that every element of $\mathcal{P}$ has an inverse in $\mathcal{A}$. If $\mathcal{P}^* \subset \mathcal{A}$ is the set of inverses of all elements* *of* $\mathcal{P}$, *then* $\mathcal{A} = \mathcal{P} + \mathcal{P}^*$, *which can be expressed*

$$\mathcal{A} = \left\{ f_c + \sum_{k=1}^{K} \alpha_k \delta_{\lambda_k} \ \middle| \ \begin{array}{l} \int_{\lambda_{\min}}^{\lambda} f_c(\mu) \, d\mu \text{ is absolutely continuous,} \\ \lambda, \lambda_k \in \mathcal{V}, \alpha_k \in \mathbb{R}, K \in \mathbb{N} \end{array} \right\}. \tag{2.8}$$

Proof. Since every element of $\mathcal{P}$ is cancellable, the fact that the inverse completion $\mathcal{A}$ has the form $\mathcal{P} + \mathcal{P}^*$ follows directly from Theorems 20.1 and 20.2 in [57] which apply to this situation. It is easy to see that any group containing $\mathcal{P}$ must also contain $\mathcal{P}^*$ and thus $\mathcal{P} + \mathcal{P}^*$ (which is a group), so this must be the smallest group containing $\mathcal{P}$. Thus any element of $\mathcal{A}$ can be written

$$f = f_1 - f_2, \qquad f_1, f_2 \in \mathcal{P}. \tag{2.9}$$

If we temporarily denote the set in Eq. (2.8) as $\mathcal{A}'$, it is clear that any element of the form of Eq. (2.9) belongs to $\mathcal{A}'$, where we have simply removed the constraint that $f_c(\lambda) \geq 0$ and $\alpha_k > 0$. This follows since from elementary properties of continuity, the sum of any two absolutely continuous functions is absolutely continuous. Thus $\mathcal{P} + \mathcal{P}^* \subset \mathcal{A}'$. Conversely, any element of $\mathcal{A}'$ can be written

$$\left(f_c^+ + \sum_{k \in \mathcal{I}_+} \alpha_k \delta_{\lambda_k} \right) - \left(f_c^- + \sum_{k \in \mathcal{I}_-} \alpha_k \delta_{\lambda_k} \right) \tag{2.10}$$

where

$$f_c^+(\lambda) = \begin{cases} f_c(\lambda) & \text{if } f_c(\lambda) \geq 0 \\ 0 & \text{if } f_c(\lambda) < 0, \end{cases} \tag{2.11}$$

$$f_c^-(\lambda) = \begin{cases} 0 & \text{if } f_c(\lambda) \geq 0 \\ -f_c(\lambda) & \text{if } f_c(\lambda) < 0, \end{cases} \tag{2.12}$$

and $\mathcal{I}_+ = \{k | \alpha_k > 0\}, \mathcal{I}_- = \{k | \alpha_k < 0\}$. This clearly belongs to $\mathcal{P} + \mathcal{P}^*$, and thus we conclude that $\mathcal{A}' = \mathcal{A}$. $\qquad \square$

Besides being able to add and subtract the elements of $\mathcal{A}$, we can also multiply them by any real scalar to get another element of $\mathcal{A}$. With this operation of scalar multiplication, the group $\mathcal{A}$ becomes a real vector space. The definition and properties of vector spaces are given in Appendix C. It is straightforward to show that $\mathcal{A}$, with the ordinary definitions of addition and scalar multiplication, satisfies all the axioms of a vector space over the field of real numbers. We henceforth use the symbol $\mathcal{A}$ to denote this vector space. We can naturally decompose $\mathcal{A}$ as a direct sum

$$\mathcal{A} = \mathcal{A}_c \oplus \mathcal{A}_d \tag{2.13}$$

where the subspaces $\mathcal{A}_c$ and $\mathcal{A}_d$ are defined in the obvious way:

$$\mathcal{A}_c = \left\{ f_c \mid \int_{\lambda_{\min}}^{\lambda} f_c(\mu)\, d\mu \text{ is absolutely continuous, } \lambda \in \mathcal{V} \right\} \tag{2.14}$$

$$\mathcal{A}_d = \left\{ \sum_{k=1}^{K} \alpha_k \delta_{\lambda_k} \mid \lambda, \lambda_k \in \mathcal{V}, \alpha_k \in \mathbb{R}, K \in \mathbb{N} \right\}. \tag{2.15}$$

This means that any $f \in \mathcal{A}$ can be written in a *unique* fashion as $f = f_c + f_d$ where $f_c \in \mathcal{A}_c$ and $f_d \in \mathcal{A}_d$. See Appendix C for the definition of direct sum, which in particular does not involve any concept of orthogonality, which is not defined in $\mathcal{A}$. Note that $\mathcal{A}_c$ is the well-known space $L^1(\mathcal{V})$ [33].

The scalar multiplication operation can be restricted to the subset $\mathcal{P}$ as long as the scalar multipliers are non-negative, i.e., $\mathcal{P}$ is closed under multiplication by non-negative scalars. The subset $\mathcal{P}$ has the form of a *convex cone* within $\mathcal{A}$. It is a cone since for any $f \in \mathcal{P}$, $\alpha f \in \mathcal{P}$ for all $\alpha > 0$ ([20], pg. 176). It is convex since, for any f_1 and f_2 belonging to $\mathcal{P}$, the convex combination $\alpha f_1 + (1 - \alpha) f_2$ also belongs to $\mathcal{P}$ for all $0 \leq \alpha \leq 1$. We can decompose $\mathcal{P}$ in the same way as $\mathcal{A}$, i.e., $\mathcal{P} = \mathcal{P}_c \oplus \mathcal{P}_d$ in a straightforward way.

There are other ways to arrive at the vector space $\mathcal{A}$ that the reader may prefer. One way is to follow the same approach usually used to obtain the integers $\mathbb{Z}$ from the natural numbers $\mathbb{N}$, namely as a quotient of $\mathcal{P} \times \mathcal{P}$ with respect to a certain equivalence relation. This approach was taken by Krantz [26]. Another approach is to consider $\mathcal{P}$ to be a subset of a suitable general vector space such as a set of distributions with support on $\mathcal{V}$. Then $\mathcal{A}$ is the subspace of this vector space generated by $\mathcal{P}$. The advantage of the approach taken above is that the exact form of elements of $\mathcal{A}$ is made evident.

2.4   METRIC ON $\mathcal{A}$

It may be useful to define a metric on $\mathcal{A}$ to be able to talk about closeness of elements of $\mathcal{A}$ or the topology of $\mathcal{A}$ in general. The natural metric for $\mathcal{A}$ is the L^1 metric. According to this metric, the distance between two elements of $\mathcal{A}$ is given by

$$d(f_1, f_2) = \int_{\mathcal{V}} |f_1(\lambda) - f_2(\lambda)|\, d\lambda, \tag{2.16}$$

which will be finite for all pairs of elements of $\mathcal{A}$. With this metric, we can establish a reasonable topology for $\mathcal{A}$. Note that the L^2 distance is *not* a suitable metric for $\mathcal{A}$, since the elements of $\mathcal{A}$ represent power, and the L^2 distance is not in general finite for elements of $\mathcal{A}_d$.

It is important to note that the metric defined here quantifies the raw difference in spectral densities, and in no way quantifies the perceptual difference between the lights as seen by a human observer.

2.5 DISCRETE REPRESENTATION OF ELEMENTS OF $\mathcal{A}$

It is frequently required to have a discrete representation of elements of $\mathcal{A}$, either to carry out computations concerning these spectral densities on a digital computer or in digital instrumentation. Because of the potential presence of Dirac deltas in elements of $\mathcal{A}$, we cannot simply apply a sampled representation and expect to get arbitrarily good accuracy. However, by separating a signal f into its components in $\mathcal{A}_c$ and $\mathcal{A}_d$ as $f = f_c + f_d$, a discrete representation is possible with excellent accuracy. It has been well established that for the purpose of color perception, elements of $\mathcal{A}_c$ can be accurately represented with sampled versions using a sample of spacing of 5 to 10 nm [52], [31], depending on the characteristics of the light sources involved. The elements of $\mathcal{A}_d$ can be represented by a set of ordered pairs, $\{(\alpha_i, \lambda_k), k = 1, \ldots, K\}$. Thus the component in $\mathcal{A}_d$ can be well represented with these $2K$ numbers. Taken together, a good discrete representation for arbitrary elements of $\mathcal{A}$ is possible for digital computations. Of course, all numerical examples in this lecture are carried out using such discrete representations.

CHAPTER 3

The Color Vector Space

3.1 INTRODUCTION

It is a *psychophysical* observation that markedly different spectral densities can give rise to identical color sensations in a human viewer. There are physiological explanations for this phenomenon based on the spectral sensitivity of the cone receptors in the human retina, but the classical theory of colorimetry is based strictly on psychophysical observations. If two spectral densities, $C_1(\lambda)$ and $C_2(\lambda)$, produce an identical color sensation for a human viewer under the same viewing conditions, they are said to form a *metameric pair*. In this lecture, we denote this by

$$C_1(\lambda) \triangle C_2(\lambda). \tag{3.1}$$

The two spectral densities are assumed to be physically realizable elements of $\mathcal{P}$ defined in Eq. (2.3). Eq. (3.1) applies to the two spectral densities as a whole, and not to a particular wavelength. It may be more correct to write $C_1 \triangle C_2, C_1, C_2 \in \mathcal{P}$, but I have used the notation of Eq. (3.1) throughout this lecture to emphasize that C_1 and C_2 are functions of wavelength.

A typical scenario to test this is shown in Fig. 3.1. Two lights with spectral densities $C_1(\lambda)$ and $C_2(\lambda)$ are displayed in the two halves of a circular target, shown on a neutral background. The diameter of the circle is typically $2°$ or $10°$, measured in angle subtended at the viewer's eye. If the circle appears to be perfectly uniform to a normal human viewer and the boundary between the two semicircular halves cannot be detected, even though $C_1(\lambda)$ and $C_2(\lambda)$ are different functions, $C_1(\lambda)$ and $C_2(\lambda)$ form a metameric pair. A typical example is illustrated in Fig. 3.2, which shows the spectral densities (in arbitrary units) of a medium pink light and an identical pink color displayed on a CRT display. This condition of metamerism applies to a *specific human observer* and will not, in general, be the same for other human observers; it may be very different for other seeing entities, such as birds, robots, cameras, etc. However, it is very *similar* for all humans with 'normal' color vision, and so we will assume the existence of a 'standard observer'.

Empirical psychophysical observations about such metameric equivalence for humans with normal color vision allow us to establish a strong mathematical structure for color measurement. We will present these observations as needed and develop the consequences. These properties are sometimes called Grassmann's laws, and so we label them G1, G2, etc. This numbering does not necessarily correspond to any other numbering of Grassmann's laws in the literature. These properties have been established empirically over more than a century and hold over a wide range of intensity, and in particular under normal viewing conditions. In the following, we assume that they hold unconditionally and take them as axioms.

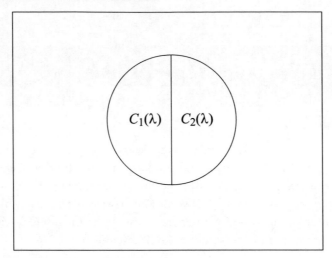

Figure 3.1: Illustration of bipartite target to test for metamerism.

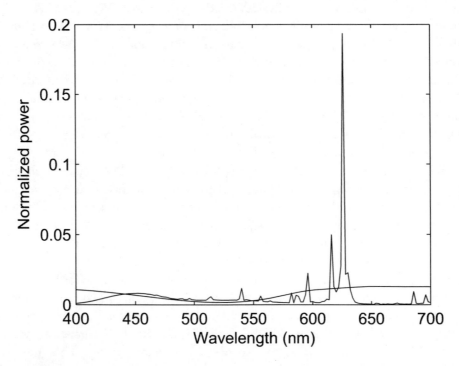

Figure 3.2: Illustration of metamers. Spectral density of a medium pink light (smooth curve) and spectral of an identical pink color displayed on a CRT (peaky curve).

3.2   PROPERTIES OF METAMERISM

Metamerism, as expressed by Eq. (3.1), is a *relation* on the set $\mathcal{P}$. The first empirical observation is transitivity:

> **Transitivity (G1).** If $C_1(\lambda) \vartriangle C_2(\lambda)$ and $C_2(\lambda) \vartriangle C_3(\lambda)$, then $C_1(\lambda) \vartriangle C_3(\lambda)$.

If this is coupled with the obvious facts that $C(\lambda) \vartriangle C(\lambda)$ and that $C_1(\lambda) \vartriangle C_2(\lambda)$ implies $C_2(\lambda) \vartriangle C_1(\lambda)$, we conclude that metamerism is an *equivalence relation* on $\mathcal{P}$ (see Appendix B).

For any $C(\lambda) \in \mathcal{P}$, we denote

$$[\mathbf{C}] = \{C_1(\lambda) \in \mathcal{P} \mid C_1(\lambda) \vartriangle C(\lambda)\}, \tag{3.2}$$

the set of all spectral densities metamerically equivalent to $C(\lambda)$; this is an equivalence class. Metameric classes are either identical or disjoint, and form a partition of $\mathcal{P}$ (Appendix B). Some of these classes may contain a single element of $\mathcal{P}$, such as 0, while most will contain an infinite number of elements. We refer to $[\mathbf{C}]$ as a *color*. Sometimes, we write $[C(\lambda)]$ to indicate the metameric equivalence class containing the specific spectral density $C(\lambda)$; any element of the equivalence class can be used as a class representative. Also, if it is important to identify the relation generating the equivalence class, we write $[\mathbf{C}]_\vartriangle$. Equality of colors is denoted with an ordinary = sign, $[\mathbf{C}_1] = [\mathbf{C}_2]$, meaning that the equivalence classes are identical. There is no universally adopted notation to represent abstract colors or metameric equivalence. For example, Wyszecki and Stiles [61] use bold capital letters, $\mathbf{C}$, Lee [30] uses bold letters in parentheses $(\mathbf{C})$ and Fairchild [12] uses bold script letters $\mathcal{C}$. I have adopted Lee's notation but with square brackets since parentheses are often used for grouping expressions, and this would lead to confusion in some of the developments that follow.

The following two empirical facts show that we can scale and add colors.

> **Scaling (G2).** If $C_1(\lambda) \vartriangle C_2(\lambda)$ then $\alpha C_1(\lambda) \vartriangle \alpha C_2(\lambda)$ for any real, nonnegative α.

Property G2 means that if $[C(\lambda)] = \{C_1(\lambda) \in \mathcal{P} \mid C_1(\lambda) \vartriangle C(\lambda)\}$ then $[\alpha C(\lambda)] = \{\alpha C_1(\lambda) \in \mathcal{P} \mid C_1(\lambda) \vartriangle C(\lambda)\}$. We denote this color $\alpha[\mathbf{C}]$ without ambiguity.

> **Addition (G3).** $C(\lambda) + C_1(\lambda) \vartriangle C(\lambda) + C_2(\lambda)$ for arbitrary $C(\lambda)$ if and only if $C_1(\lambda) \vartriangle C_2(\lambda)$.

Corollary to G3. *If* $C_1(\lambda) \vartriangle C_2(\lambda)$, *then* $C_1(\lambda) + C_3(\lambda) \vartriangle C_2(\lambda) + C_4(\lambda)$ *if and only if* $C_3(\lambda) \vartriangle C_4(\lambda)$.

Proof. Apply G3 twice and use transitivity. By G3, $C_1(\lambda) + C_3(\lambda) \vartriangle C_2(\lambda) + C_3(\lambda)$, and again by G3, $C_2(\lambda) + C_3(\lambda) \vartriangle C_2(\lambda) + C_4(\lambda)$ if and only if $C_3(\lambda) \vartriangle C_4(\lambda)$. Transitivity (G1) establishes the result. □

Property G3 allows us to define addition of colors. Let $[\mathbf{C}_1]$ and $[\mathbf{C}_2]$ be two colors. Define the sum of $[\mathbf{C}_1]$ and $[\mathbf{C}_2]$ to be

$$[\mathbf{C}_1] + [\mathbf{C}_2] = [C_1(\lambda) + C_2(\lambda)] \tag{3.3}$$

where $C_1(\lambda)$ is any element of the class $[\mathbf{C}_1]$ and $C_2(\lambda)$ is any element of the class $[\mathbf{C}_2]$. This assignment is well-defined if it is independent of the particular choices $C_1(\lambda)$ and $C_2(\lambda)$ within $[\mathbf{C}_1]$ and $[\mathbf{C}_2]$, respectively. Indeed, let $C_{11}(\lambda)$ and $C_{12}(\lambda)$ be two arbitrary elements of $[\mathbf{C}_1]$ and $C_{21}(\lambda)$ and $C_{22}(\lambda)$ be two arbitrary elements of $[\mathbf{C}_2]$. Then

$$C_{11}(\lambda) + C_{21}(\lambda) \triangleq C_{11}(\lambda) + C_{22}(\lambda)$$
$$\triangleq C_{12}(\lambda) + C_{22}(\lambda)$$

by applying G3 twice. Thus addition of colors as defined by equation (3.3) is well defined, and it is clearly commutative.

We denote the set of colors just defined $\mathcal{C}_P$. Based on the properties established above, $(\mathcal{C}_P, +)$ has an algebraic structure already introduced: a commutative semigroup with a neutral element $[0]$ in which every element is cancellable.

3.3   EXTENSION OF METAMERIC PROPERTIES TO $\mathcal{A}$

As with the space of physical stimuli $\mathcal{P}$, the space of colors $\mathcal{C}_P$ allows for addition and multiplication by positive real constants. We want to embed $\mathcal{C}_P$ in a vector space so that the powerful tools of linear algebra can be applied. A perfectly valid solution is to embed $\mathcal{C}_P$ in its inverse completion, as was done to embed $\mathcal{P}$ in $\mathcal{A}$. This is essentially what will be done. However, to maintain the identity of the space as a set of equivalence classes, the equivalence relation on $\mathcal{P}$ will be extended to all of $\mathcal{A}$, and the color space will be the set of equivalence classes of this extended relation on $\mathcal{A}$. Any element of $\mathcal{A}$ can be written as $C_a(\lambda) - C_b(\lambda)$ where $C_a(\lambda), C_b(\lambda) \in \mathcal{P}$; this representation is not unique. Intuitively, based on how color matches are carried out in practice, we would like to say that two arbitrary elements $C_1(\lambda)$ and $C_2(\lambda)$ of $\mathcal{A}$ are equivalent if and only if

$$C_{1a}(\lambda) + C_{2b}(\lambda) \triangleq C_{2a}(\lambda) + C_{1b}(\lambda). \tag{3.4}$$

Since both $C_{1a}(\lambda) + C_{2b}(\lambda)$ and $C_{2a}(\lambda) + C_{1b}(\lambda)$ are elements of $\mathcal{P}$, such metameric equivalence is physically meaningful and can be experimentally tested. However, this would have to hold no matter how $C_1(\lambda)$ and $C_2(\lambda)$ are decomposed as the difference of elements of $\mathcal{P}$. We introduce a specific decomposition and show that it extends metameric equivalence to $\mathcal{A}$ and then show that the property of equation (3.4) holds for any decomposition.

For an arbitrary $C(\lambda) \in \mathcal{A}$, define the following elements of $\mathcal{P}$ as was done in Section 2.3:

$$C^+(\lambda) = \begin{cases} C(\lambda) & \text{if } C(\lambda) \geq 0, \\ 0 & \text{if } C(\lambda) < 0. \end{cases}$$

$$C^-(\lambda) = \begin{cases} 0 & \text{if } C(\lambda) \geq 0, \\ -C(\lambda) & \text{if } C(\lambda) < 0. \end{cases}$$

Both $C^+(\lambda) \in \mathcal{P}$ and $C^-(\lambda) \in \mathcal{P}$, and $C(\lambda) = C^+(\lambda) - C^-(\lambda)$. Moreover, if $C(\lambda) \in \mathcal{P}$, then $C^+(\lambda) = C(\lambda)$ and $C^-(\lambda) = 0$. The presence of possible Dirac deltas is dealt with informally here. The more formal approach of Section 2.3 can be applied if desired. We now define the following relation on $\mathcal{A}$:

Definition 3.1

$$C_1(\lambda) \boxminus C_2(\lambda) \quad \text{if and only if} \quad C_1^+(\lambda) + C_2^-(\lambda) \triangleq C_2^+(\lambda) + C_1^-(\lambda). \tag{3.5}$$

Since both $C_1^+(\lambda) + C_2^-(\lambda)$ and $C_2^+(\lambda) + C_1^-(\lambda)$ are elements of $\mathcal{P}$, we can physically test for

metameric equivalence. Note that if $C_1(\lambda) \in \mathcal{P}$ and $C_2(\lambda) \in \mathcal{P}$, then $C_1(\lambda) \boxminus C_2(\lambda)$ if and only if $C_1(\lambda) \triangleq C_2(\lambda)$, so that this relation is indeed an *extension* of metameric equivalence on $\mathcal{P}$ to a relation on $\mathcal{A}$. In fact, we will show that $\boxminus$ is also an equivalence relation and that Grassmann's laws can be suitably extended (denoted Gi'). To avoid obscuring the logical flow, some longer proofs have been collected at the end of this section.

Proposition 3.2   *The relation $\boxminus$ on $\mathcal{A}$ is an equivalence relation.*

Proof. We must show that reflexivity, symmetry and transitivity of $\triangleq$ imply reflexivity, symmetry and transitivity of $\boxminus$. See Section 3.3.1 for the proof.   □

To turn the set of equivalence classes into a vector space, we must also verify that properties G2 and G3 continue to hold for the relation $\boxminus$.

Theorem 3.3   *If $C_1(\lambda) \boxminus C_2(\lambda)$, then $\alpha C_1(\lambda) \boxminus \alpha C_2(\lambda)$, $\forall \alpha \in \mathbb{R}$ (G2')*

Proof. See Section 3.3.1.   □

Theorem 3.4 $C(\lambda) + C_1(\lambda) \boxminus C(\lambda) + C_2(\lambda)$ *for arbitrary $C(\lambda) \in \mathcal{A}$ if and only if $C_1(\lambda) \boxminus C_2(\lambda)$ (G3').*

This is not an obvious result since the regions where $C(\lambda) + C_1(\lambda)$ is positive and negative do not coincide with those of $C(\lambda)$ or $C_1(\lambda)$, and similarly for $C(\lambda) + C_2(\lambda)$. We first establish the following simpler results.

Proposition 3.5   *Let $C_1(\lambda), C_2(\lambda) \in \mathcal{P}$. Then $C_1(\lambda) - C_2(\lambda) \boxminus 0$ if and only if $C_1(\lambda) \triangleq C_2(\lambda)$.*

Proof. See Section 3.3.1.   □

Proposition 3.6   *Let $C_1(\lambda), C_2(\lambda) \in \mathcal{A}$. Then*

$$C_1(\lambda) \boxminus C_2(\lambda) \quad \text{if and only if} \quad C_1(\lambda) - C_2(\lambda) \boxminus 0$$

Proof. By Proposition 3.5, $C_1^+(\lambda) + C_2^-(\lambda) \mathbin{\triangle} C_2^+(\lambda) + C_1^-(\lambda)$ if and only if $(C_1^+(\lambda) + C_2^-(\lambda)) - (C_2^+(\lambda) + C_1^-(\lambda)) \boxminus 0$, i.e., $C_1(\lambda) - C_2(\lambda) \boxminus 0$. □

Corollary to Proposition 3.6: $C_1(\lambda) \boxminus C_2(\lambda)$ *if and only if* $C_1(\lambda) = C_2(\lambda) + C_0(\lambda)$ *where* $C_0(\lambda) \boxminus 0$.

Proof. $C_1(\lambda) - C_2(\lambda) \boxminus 0$ means $C_1(\lambda) - C_2(\lambda) \in [0]$. Thus, denoting $C_0(\lambda) = C_1(\lambda) - C_2(\lambda)$, the result follows. □

With these propositions, the proof of Theorem 3.4 is now straightforward.

Proof of Theorem 3.4: By Proposition 3.6, $C(\lambda) + C_1(\lambda) \boxminus C(\lambda) + C_2(\lambda)$ if and only if $(C(\lambda) + C_1(\lambda)) - (C(\lambda) + C_2(\lambda)) \boxminus 0$, i.e., $C_1(\lambda) - C_2(\lambda) \boxminus 0$, which again by Proposition 3.6 occurs if and only if $C_1(\lambda) \boxminus C_2(\lambda)$. □

We can also use Proposition 3.6 to show that the extended definition of equivalence satisfies the intuitive property of equation (3.4).

Theorem 3.7   *Suppose that $C_1(\lambda) = C_{1a}(\lambda) - C_{1b}(\lambda)$ and $C_2(\lambda) = C_{2a}(\lambda) - C_{2b}(\lambda)$ where $C_{1a}(\lambda), C_{1b}(\lambda), C_{2a}(\lambda), C_{2b}(\lambda) \in \mathcal{P}$. Then $C_1(\lambda) \boxminus C_2(\lambda)$ if and only if $C_{1a}(\lambda) + C_{2b}(\lambda) \mathbin{\triangle} C_{2a}(\lambda) + C_{1b}(\lambda)$.*

Proof. By Proposition 3.6 $C_1(\lambda) \boxminus C_2(\lambda)$ if and only if $(C_{1a}(\lambda) - C_{1b}(\lambda)) - (C_{2a}(\lambda) - C_{2b}(\lambda)) \boxminus 0$, or equivalently, if and only if $(C_{1a}(\lambda) + C_{2b}(\lambda)) - (C_{2a}(\lambda) + C_{1b}(\lambda)) \boxminus 0$. Applying Proposition 3.6 again, this hold if and only if $C_{1a}(\lambda) + C_{2b}(\lambda) \boxminus C_{2a}(\lambda) + C_{1b}(\lambda)$. But since both $C_{1a}(\lambda) + C_{2b}(\lambda)$ and $C_{2a}(\lambda) + C_{1b}(\lambda)$ are elements of $\mathcal{P}$, this is true if and only if $C_{1a}(\lambda) + C_{2b}(\lambda) \mathbin{\triangle} C_{2a}(\lambda) + C_{1b}(\lambda)$. □

3.3.1   PROOFS OF PROPOSITIONS AND THEOREMS OF SECTION 3.3

Proposition 3.2: *The relation $\boxminus$ on $\mathcal{A}$ is an equivalence relation.*

Proof. *reflexivity:* $C^+(\lambda) + C^-(\lambda) \mathbin{\triangle} C^+(\lambda) + C^-(\lambda)$ by reflexivity of $\triangle$, and thus $C(\lambda) \boxminus C(\lambda)$.
symmetry: If $C_1(\lambda) \boxminus C_2(\lambda)$ then $C_1^+(\lambda) + C_2^-(\lambda) \mathbin{\triangle} C_2^+(\lambda) + C_1^-(\lambda)$. By symmetry of $\triangle$, $C_2^+(\lambda) + C_1^-(\lambda) \mathbin{\triangle} C_1^+(\lambda) + C_2^-(\lambda)$ and thus $C_2(\lambda) \boxminus C_1(\lambda)$.
transitivity (G1'): Suppose that $C_1(\lambda) \boxminus C_2(\lambda)$ and $C_2(\lambda) \boxminus C_3(\lambda)$. Then

$$C_1^+(\lambda) + C_2^-(\lambda) \mathbin{\triangle} C_2^+(\lambda) + C_1^-(\lambda) \quad \text{and}$$
$$C_2^+(\lambda) + C_3^-(\lambda) \mathbin{\triangle} C_3^+(\lambda) + C_2^-(\lambda).$$

Applying G3,

$$C_1^+(\lambda) + C_2^-(\lambda) + C_3^-(\lambda) \triangleq C_2^+(\lambda) + C_1^-(\lambda) + C_3^-(\lambda) \quad \text{and}$$
$$C_2^+(\lambda) + C_3^-(\lambda) + C_1^-(\lambda) \triangleq C_3^+(\lambda) + C_2^-(\lambda) + C_1^-(\lambda).$$

Applying transitivity of $\triangleq$, $C_1^+(\lambda) + C_2^-(\lambda) + C_3^-(\lambda) \triangleq C_3^+(\lambda) + C_2^-(\lambda) + C_1^-(\lambda)$, and applying G3 again, $C_1^+(\lambda) + C_3^-(\lambda) \triangleq C_3^+(\lambda) + C_1^-(\lambda)$, i.e., $C_1(\lambda) \boxminus C_3(\lambda)$. $\qquad\square$

Theorem 3.3 *If $C_1(\lambda) \boxminus C_2(\lambda)$, then $\alpha C_1(\lambda) \boxminus \alpha C_2(\lambda)$, $\forall \alpha \in \mathbb{R}$ (G2′).*

Proof. Consider first the case $\alpha \geq 0$. By hypothesis, $C_1^+(\lambda) + C_2^-(\lambda) \triangleq C_2^+(\lambda) + C_1^-(\lambda)$. Applying G2, $\alpha(C_1^+(\lambda) + C_2^-(\lambda)) \triangleq \alpha(C_2^+(\lambda) + C_1^-(\lambda))$, i.e. $\alpha C_1^+(\lambda) + \alpha C_2^-(\lambda) \triangleq \alpha C_2^+(\lambda) + \alpha C_1^-(\lambda)$. Now, for $\alpha \geq 0$, it follows that $(\alpha C)^+ = \alpha C^+$ and $(\alpha C)^- = \alpha C^-$, and so

$$(\alpha C_1)^+(\lambda) + (\alpha C_2)^-(\lambda) \triangleq (\alpha C_2)^+(\lambda) + (\alpha C_1)^-(\lambda), \text{ i.e.,}$$

$$\alpha C_1(\lambda) \boxminus \alpha C_2(\lambda).$$

Now consider the case $\alpha < 0$. Then,

$$(\alpha C)^+ = |\alpha| C^- = -\alpha C^- \quad \text{and}$$
$$(\alpha C)^- = |\alpha| C^+ = -\alpha C^+.$$

Now applying G2 with $-\alpha$,

$$-\alpha(C_1^+(\lambda) + C_2^-(\lambda)) \triangleq -\alpha(C_2^+(\lambda) + C_1^-(\lambda))$$

$$(\alpha C_1)^-(\lambda) + (\alpha C_2)^+(\lambda) \triangleq (\alpha C_2)^-(\lambda) + (\alpha C_1)^+(\lambda)$$

Rearranging,

$$(\alpha C_1)^+(\lambda) + (\alpha C_2)^-(\lambda) \triangleq (\alpha C_2)^+(\lambda) + (\alpha C_1)^-(\lambda), \text{ i.e.,}$$

$$\alpha C_1(\lambda) \boxminus \alpha C_2(\lambda)$$

$\qquad\square$

Proposition 3.5: *Let $C_1(\lambda), C_2(\lambda) \in \mathcal{P}$. Then $C_1(\lambda) - C_2(\lambda) \boxminus 0$ if and only if $C_1(\lambda) \triangleq C_2(\lambda)$.*

Proof. Let $\mathcal{I}_1 = \{\lambda \subset \mathcal{V} \mid C_1(\lambda) \geq C_2(\lambda)\}$ and $\mathcal{I}_2 = \{\lambda \subset \mathcal{V} \mid C_1(\lambda) < C_2(\lambda)\}$. Then $\mathcal{I}_1 \cap \mathcal{I}_2 = \emptyset$ and $\mathcal{I}_1 \cup \mathcal{I}_2 = \mathcal{V}$. Thus

$$(C_1(\lambda) - C_2(\lambda))^+ = \begin{cases} C_1(\lambda) - C_2(\lambda) & \lambda \in \mathcal{I}_1 \\ 0 & \lambda \in \mathcal{I}_2 \end{cases}$$

and

$$(C_1(\lambda) - C_2(\lambda))^- = \begin{cases} 0 & \lambda \in \mathcal{I}_1 \\ C_2(\lambda) - C_1(\lambda) & \lambda \in \mathcal{I}_2 \end{cases}$$

Note that

$$(C_1(\lambda) - C_2(\lambda))^+ + C_2(\lambda) = \begin{cases} C_1(\lambda) & \lambda \in \mathcal{I}_1 \\ C_2(\lambda) & \lambda \in \mathcal{I}_2 \end{cases}$$

and

$$(C_1(\lambda) - C_2(\lambda))^- + C_1(\lambda) = \begin{cases} C_1(\lambda) & \lambda \in \mathcal{I}_1 \\ C_2(\lambda) & \lambda \in \mathcal{I}_2 \end{cases}$$

Thus trivially

$$(C_1(\lambda) - C_2(\lambda))^+ + C_2(\lambda) \triangle (C_1(\lambda) - C_2(\lambda))^- + C_1(\lambda).$$

It follows from the corollary to G3 that if $(C_1(\lambda) - C_2(\lambda))^+ \triangle (C_1(\lambda) - C_2(\lambda))^-$, i.e., $C_1(\lambda) - C_2(\lambda) \boxminus 0$, then $C_1(\lambda) \triangle C_2(\lambda)$ and conversely, if $C_1(\lambda) \triangle C_2(\lambda)$ then $(C_1(\lambda) - C_2(\lambda))^+ \triangle (C_1(\lambda) - C_2(\lambda))^-$. □

3.4   DEFINITION AND PROPERTIES OF THE COLOR VECTOR SPACE

We are now in a position to define the color vector space.

Theorem 3.8   *The set of equivalence classes of $\boxminus$ on the vector space $\mathcal{A}$ forms a real vector space that we call the* color vector space $\mathcal{C} = \{[C(\lambda)]_\boxminus \mid C(\lambda) \in \mathcal{A}\}$.

Proof. According to Theorem 3.3 and Theorem 3.4, addition and multiplication by a real scalar are well-defined operations on the set of equivalence classes of $\boxminus$. The neutral element is $[0]_\boxminus$, and the negative of any element is $-[C(\lambda)]_\boxminus = [-C(\lambda)]_\boxminus$. The other requisite properties of a vector space (see Appendix C) follow simply from the properties of real numbers. □

In the following sections, we will always assume that the equivalence relation is $\boxminus$ and write simply $[\cdot]$ without the subscript to denote an element of $\mathcal{C}$.

I am sure that the reader strongly suspects that the color vector space $\mathcal{C}$ is a finite-dimensional vector space of dimension 3. The next empirical observation allows us to establish the dimension of $\mathcal{C}$, for trichromats.

> **Dimension (G4).** Four physically realizable colors are always linearly dependent in the following sense: For $C(\lambda), C_1(\lambda), C_2(\lambda), C_3(\lambda)$ arbitrary elements of $\mathcal{P}$, there always exist $\alpha \geq 0$ and non-negative $\alpha_1, \alpha_2, \alpha_3$, not all zero, such that
>
> $$\alpha C(\lambda) + \sum_{i \in I} \alpha_i C_i(\lambda) \triangleq \sum_{i \in \overline{I}} \alpha_i C_i(\lambda)$$
>
> where $I \subset \{1, 2, 3\}$ and $\overline{I} = \{1, 2, 3\} \backslash I$. This is *not* the case for three colors.

In other words, either $C(\lambda)$ matches a weighted combination of the three $C_i(\lambda)$, *or* a weighted combination of $C(\lambda)$ and one of the $C_i(\lambda)$ matches a weighted combination of the other two, *or* a weighted combination of $C(\lambda)$ and two of the $C_i(\lambda)$ matches the third. This result allows us to determine the dimension of the color space $\mathcal{C}$.

Theorem 3.9   *The dimension of $\mathcal{C}$ is 3.*

Proof. Using G4 and previous results, we must show that we can find three linearly independent colors that span $\mathcal{C}$. Property G4 implies that we can find three elements of $\mathcal{P}$, $P_1(\lambda), P_2(\lambda), P_3(\lambda)$ such that no one of them can match a linear combination of the other two. It follows that $[P_1(\lambda)]$, $[P_2(\lambda)]$, and $[P_3(\lambda)]$ are linearly independent in $\mathcal{C}$. Otherwise, there would exist $\alpha_1, \alpha_2, \alpha_3$, not all zero, such that $\alpha_1[P_1(\lambda)] + \alpha_2[P_2(\lambda)] + \alpha_3[P_3(\lambda)] \boxminus [0]$, which contradicts the above statement. We need only prove that $[\mathbf{P}_1], [\mathbf{P}_2]$ and $[\mathbf{P}_3]$ span $\mathcal{C}$ to establish the proof. Let $C(\lambda)$ be an element of $\mathcal{A}$. If $C(\lambda) \in \mathcal{P}$ or $-C(\lambda) \in \mathcal{P}$, it is clear from G4 that we can find $\alpha_1, \alpha_2, \alpha_3$ such that $C(\lambda) \boxminus \alpha_1 P_1(\lambda) + \alpha_2 P_2(\lambda) + \alpha_3 P_3(\lambda)$. Now suppose that neither of the above two conditions hold and that $C(\lambda)$ is an arbitrary element of $\mathcal{C}$. Thus $C(\lambda) = C^+(\lambda) - C^-(\lambda)$ where both $C^+(\lambda)$ and $C^-(\lambda)$ are nonzero elements of $\mathcal{P}$. From G4 we can find $\alpha_1^+, \alpha_2^+, \alpha_3^+$ such that

$$C^+(\lambda) \boxminus \alpha_1^+ P_1(\lambda) + \alpha_2^+ P_2(\lambda) + \alpha_3^+ P_3(\lambda),$$

and similarly, we can find $\alpha_1^-, \alpha_2^-, \alpha_3^-$ such that

$$C^-(\lambda) \boxminus \alpha_1^- P_1(\lambda) + \alpha_2^- P_2(\lambda) + \alpha_3^- P_3(\lambda).$$

It follows that

$$C^+(\lambda) - C^-(\lambda) \boxminus (\alpha_1^+ - \alpha_1^-) P_1(\lambda) + (\alpha_2^+ - \alpha_2^-) P_2(\lambda) + (\alpha_3^+ - \alpha_3^-) P_3(\lambda)$$

or in other words

$$[\mathbf{C}] = (\alpha_1^+ - \alpha_1^-)[\mathbf{P}_1] + (\alpha_2^+ - \alpha_2^-)[\mathbf{P}_2] + (\alpha_3^+ - \alpha_3^-)[\mathbf{P}_3].$$

Thus $[\mathbf{P}_1], [\mathbf{P}_2]$ and $[\mathbf{P}_3]$ do indeed span $\mathcal{C}$, and so $\mathcal{C}$ must be of dimension 3. $\square$

Any three linearly independent colors $[\mathbf{P}_1]$, $[\mathbf{P}_2]$ and $[\mathbf{P}_3]$ form a basis for $\mathcal{C}$ and are referred to as *primaries*. The coefficients in the expansion of a color with respect to a set of primaries are called *tristimulus values*. For any $[\mathbf{C}] \in \mathcal{C}$ we have

$$[\mathbf{C}] = C_1[\mathbf{P}_1] + C_2[\mathbf{P}_2] + C_3[\mathbf{P}_3] \tag{3.6}$$

where the $C_i \in \mathbb{R}$ are tristimulus values. We generally use uppercase letters with an appropriate subscript related to the primaries to denote tristimulus values. Note that Eq. (3.6) defines a *representation* for a color by a set of three real numbers, the tristimulus values. It does not necessarily refer to the physical synthesis of a color, which is a separate process. The tristimulus values can be positive or negative, and the primaries do not have to be physically realizable colors. The physical synthesis of colors with real primaries is discussed in Section 3.10. As with all finite-dimensional vector spaces, it is very convenient to represent the coordinates (tristimulus values) of a given element of $\mathcal{C}$ with respect to a particular basis as a column matrix, in order to use the powerful techniques of matrix algebra. Let $\mathcal{B} = \{[\mathbf{P}_1], [\mathbf{P}_2], [\mathbf{P}_3]\}$ denote a basis for $\mathcal{C}$. Then we express the tristimulus values with respect to $\mathcal{B}$ as a column matrix

$$\mathbf{C}_\mathcal{B} = \begin{bmatrix} C_1 \\ C_2 \\ C_3 \end{bmatrix}. \tag{3.7}$$

It is important to identify the basis as virtually all color processing systems use more than one basis in their characterization.

Any color can be represented by a triple of tristimulus values with respect to a set of primaries and thus colors could be plotted on a three-dimensional set of coordinate axes. It should be emphasized that in such a diagram, usual Cartesian concepts such as distance and orthogonality are not meaningful. There is no requirement for the $[\mathbf{P}_i]$ to be physical colors, and in fact, they are not for several widely used color systems.

By convention (but not always), the intensities of primaries are chosen so that the sum of one unit of each matches a selected reference white, e.g., equal energy white:

$$[\mathbf{P}_1] + [\mathbf{P}_2] + [\mathbf{P}_3] = [\mathbf{W}]. \tag{3.8}$$

Thus, if $[\mathbf{P}_1']$, $[\mathbf{P}_2']$ and $[\mathbf{P}_3']$ are linearly independent elements of $\mathcal{C}$ and $[\mathbf{W}]$ is the selected reference white, we would get the unique representation of $[\mathbf{W}]$

$$[\mathbf{W}] = W_1[\mathbf{P}_1'] + W_2[\mathbf{P}_2'] + W_3[\mathbf{P}_3'].$$

We then choose as primaries $[\mathbf{P}_i] = W_i[\mathbf{P}_i']$, $i = 1, 2, 3$, which satisfy the convention. It is not necessary for primaries to satisfy this condition. For example, if reference white is selected as one of the primaries (which is certainly possible), then the convention would not make sense.

An example of a set of primaries would be the equivalence classes containing the spectral densities of the light emitted by the red, green and blue phosphors used in CRT television displays

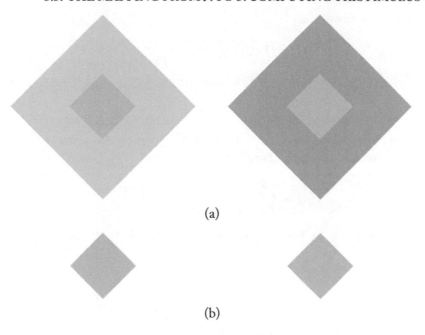

(a)

(b)

Figure 3.3:   Illustration of importance of context. The small diamonds on the left and the right have the same tristimulus values. They appear quite different in (a) when viewed in different context whereas they appear the same in (b) when viewed in the same context.

(Fig. 2.1(c)). These are indeed linearly independent as has been established empirically over the years. These will be used as a recurring example in this chapter. In general, three spectral densities chosen at random will almost certainly be linearly independent.

It is important to emphasize that the tristimulus values do not specify the perceptual appearance of a color. Two spectral densities with the same tristimulus values will appear the same when viewed in the *same* context. However, they may appear quite different when viewed in *different* contexts. This is illustrated in the striking example of Fig. 3.3.

3.5   THE MAPPING FROM $\mathcal{A}$ TO $\mathcal{C}$: COMPUTING TRISTIMULUS VALUES

We have now established that $\mathcal{A}$ and $\mathcal{C}$ are vector spaces, and that the mapping $\mathcal{S}$ that assigns to each $C(\lambda) \in \mathcal{A}$ the equivalence class $[C(\lambda)] \in \mathcal{C}$ is a vector space homomorphism. Specifically, $\mathcal{S} : \mathcal{A} \to \mathcal{C} : C(\lambda) \mapsto [C(\lambda)]$ satisfies $\mathcal{S}(\alpha_1 C_1(\lambda) + \alpha_2 C_2(\lambda)) = \alpha_1 \mathcal{S}(C_1(\lambda)) + \alpha_2 \mathcal{S}(C_2(\lambda))$. The following theorem shows the specific form that the mapping $\mathcal{S}$ takes given a particular basis for $\mathcal{C}$.

Theorem 3.10   *Let $\mathcal{B} = \{[\mathbf{P}_1], [\mathbf{P}_2], [\mathbf{P}_2]\}$ be a set of primaries forming a basis for the color space $\mathcal{C}$. There exists a set of three functions $\bar{p}_1(\lambda)$, $\bar{p}_2(\lambda)$ and $\bar{p}_3(\lambda)$ such that for any $C(\lambda) \in \mathcal{A}$, then $[C(\lambda)] =$*

$C_1[\mathbf{P}_1] + C_2[\mathbf{P}_2] + C_3[\mathbf{P}_3]$ *where*

$$C_i = \int_{\mathcal{V}} C(\lambda)\bar{p}_i(\lambda)\,d\lambda. \tag{3.9}$$

The three functions $\bar{p}_i(\lambda), i = 1, 2, 3$ *are called the* color matching functions *relative to the given primaries.*

Proof. Any element of $\mathcal{A}$ can be written $C(\lambda) = C_c(\lambda) + \sum_{k=1}^{K} \alpha_k \delta(\lambda - \lambda_k)$, where $C_c(\lambda)$ is continuous almost everywhere. Let $q_\Delta(\lambda)$ be defined as

$$q_\Delta(\lambda) = \begin{cases} \frac{1}{\Delta} & \text{if } -\frac{\Delta}{2} < \lambda \le \frac{\Delta}{2}; \\ 0 & \text{elsewhere.} \end{cases} \tag{3.10}$$

We note that $\lim_{\Delta \to 0} q_\Delta(\lambda) = \delta(\lambda)$. We can form a piecewise constant approximation of $C_c(\lambda)$ as

$$C_\Delta(\lambda) = \sum_{i=0}^{N} \Delta C_c(\mu_i) q_\Delta(\lambda - \mu_i) \tag{3.11}$$

where $\mu_i = \lambda_{\min} + i\Delta$ and $\Delta = (\lambda_{\max} - \lambda_{\min})/N$. $C_\Delta(\lambda)$ converges pointwise to $C_c(\lambda)$ for any $\lambda \in \mathcal{V}$ as $\Delta \to 0$. By the linearity of our mapping $\mathcal{S}$ from $\mathcal{A}$ to $\mathcal{C}$, we have

$$[C_\Delta(\lambda)] = \sum_{i=0}^{N} \Delta C_c(\mu_i)[q_\Delta(\lambda - \mu_i)]. \tag{3.12}$$

Note that as Δ gets small, $q_\Delta(\lambda - \mu_i)$ is approximately a monochromatic light at wavelength μ_i with unit power, i.e., $[q_\Delta(\lambda - \mu_i)] \to [\delta(\lambda - \mu_i)]$. Now, we can express the color $[q_\Delta(\lambda - \mu_i)]$ in terms of the given basis

$$[q_\Delta(\lambda - \mu_i)] = \bar{p}_{1\Delta}(\mu_i)[\mathbf{P}_1] + \bar{p}_{2\Delta}(\mu_i)[\mathbf{P}_2] + \bar{p}_{3\Delta}(\mu_i)[\mathbf{P}_3], \tag{3.13}$$

where $\bar{p}_{k\Delta}(\mu_i), k = 1, 2, 3$ denote the tristimulus values of $[q_\Delta(\lambda - \mu_i)]$. It follows that

$$[C_\Delta(\lambda)] = \left(\sum_{i=0}^{N} \Delta C_c(\mu_i)\bar{p}_{1\Delta}(\mu_i) \right) [\mathbf{P}_1] + \left(\sum_{i=0}^{N} \Delta C_c(\mu_i)\bar{p}_{2\Delta}(\mu_i) \right) [\mathbf{P}_2]$$
$$+ \left(\sum_{i=0}^{N} \Delta C_c(\mu_i)\bar{p}_{3\Delta}(\mu_i) \right) [\mathbf{P}_3]. \tag{3.14}$$

Taking the limit as $\Delta \to 0$,

$$[C_c(\lambda)] = \left(\int_{\mathcal{V}} C_c(\mu)\bar{p}_1(\mu)\,d\mu \right) [\mathbf{P}_1] + \left(\int_{\mathcal{V}} C_c(\mu)\bar{p}_2(\mu)\,d\mu \right) [\mathbf{P}_2]$$
$$+ \left(\int_{\mathcal{V}} C_c(\mu)\bar{p}_3(\mu)\,d\mu \right) [\mathbf{P}_3], \tag{3.15}$$

where $\bar{p}_k(\mu) = \lim_{\Delta \to 0} \bar{p}_{k\Delta}(\mu)$. It follows that

$$[\delta(\lambda - \mu)] = \bar{p}_1(\mu)[\mathbf{P}_1] + \bar{p}_2(\mu)[\mathbf{P}_2] + \bar{p}_3(\mu)[\mathbf{P}_3] \tag{3.16}$$

for all $\mu \in \mathcal{V}$. If we apply this to the discrete portion of $C(\lambda)$, we conclude that the tristimulus values of $C_d(\lambda)$ are $\sum_{k=1}^{K} \alpha_k \bar{p}_i(\lambda_k)$, $i = 1, 2, 3$, which is consistent with

$$[C_d(\lambda)] = \left(\int_{\mathcal{V}} C_d(\mu)\bar{p}_1(\mu)\, d\mu \right) [\mathbf{P}_1] + \left(\int_{\mathcal{V}} C_d(\mu)\bar{p}_2(\mu)\, d\mu \right) [\mathbf{P}_2]$$
$$+ \left(\int_{\mathcal{V}} C_d(\mu)\bar{p}_3(\mu)\, d\mu \right) [\mathbf{P}_3]. \tag{3.17}$$

Combining equations 3.15 and 3.17 and changing the dummy variable of integration from μ to λ completes the proof. In practice, the $\bar{p}_i$ are continuous and these limits are well behaved.   □

A more informal proof starts by defining the color matching functions directly from Eq. (3.16). By the sifting property of the Dirac delta, we can express $C_c(\lambda)$ as

$$C_c(\lambda) = \int_{\mathcal{V}} C_c(\mu)\delta(\lambda - \mu)\, d\mu \tag{3.18}$$

which can be seen as the superposition of an infinite number of Dirac deltas with infinitesmal weights $C_c(\mu)d\mu$. If we directly apply the linearity of color matching to this expression, the same result follows.

Fig. 3.4 shows the color matching functions that correspond to the Sony CRT RGB primaries mentioned above. We will see how these are obtained shortly. The color matching functions can be seen as the key defining entities of a color space. Many developments start with a set of three color matching functions and define the color space from them; this approach will be introduced later in this lecture.

3.6   BLACK SPACE AND THE CANONICAL DECOMPOSITION OF THE STIMULUS SPACE

We have seen that the linear transformation $\mathcal{S}$ maps the space $\mathcal{A}$ onto the three-dimensional color space $\mathcal{C}$. We denote the *kernel* or *null space* of $\mathcal{S}$ as

$$\mathcal{K} = \ker \mathcal{S} = \{C(\lambda) \in \mathcal{A} \mid \mathcal{S}(C(\lambda)) = [0]\}, \tag{3.19}$$

which is a subspace of $\mathcal{A}$. $\mathcal{K}$ is also called *black space* since any element of $\mathcal{K}$ is metameric to $C(\lambda) = 0$, i.e., black.

For any $C(\lambda) \in \mathcal{A}$, the set

$$C(\lambda) + \mathcal{K} = \{C(\lambda) + K(\lambda) \mid K(\lambda) \in \mathcal{K}\} \tag{3.20}$$

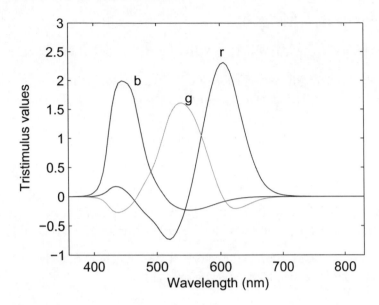

Figure 3.4: Color matching functions of the primaries determined by Sony Trinitron CRT RGB phosphors whose spectral densities are shown in Fig. 2.1(c).

is called a *coset* of $\mathcal{K}$ in $\mathcal{A}$. Proposition 3.6 basically states that $C_1(\lambda) \boxminus C_2(\lambda)$ if and only if $C_1(\lambda)$ and $C_2(\lambda)$ belong to the same coset of $\mathcal{K}$ in $\mathcal{A}$. Thus our color space is the set of cosets of $\ker \mathcal{S}$ in $\mathcal{A}$, that is the *quotient space* or *factor space* $\mathcal{A}/\mathcal{K}$. It follows that $\mathcal{A} = \mathcal{K}' \oplus \mathcal{K}$, where $\mathcal{K}'$ is isomorphic to $\mathcal{C}$, and thus is of dimension 3. However, $\mathcal{K}'$ is not unique whereas $\mathcal{K}$ is. See Appendix C for more details on quotient spaces and the resulting canonical decompositions. This is a key concept in this lecture.

For $\mathcal{K}'$, we can use the subspace of $\mathcal{A}$ spanned by any three spectral densities whose equivalence classes are linearly independent in $\mathcal{C}$, such as the spectral densities of red, green and blue lights emitted by a CRT or LCD display shown in Fig. 2.1 (c) and (d). In this case, any spectral density emitted by the display would belong to $\mathcal{K}'$. The representation of an arbitrary $C(\lambda) \in \mathcal{A}$ is straightforward. Let $P_1(\lambda), P_2(\lambda), P_3(\lambda)$ form a basis for $\mathcal{K}'$, such as one of the ones given above. As in the previous section, we can find the color matching functions $\bar{p}_1(\lambda), \bar{p}_2(\lambda), \bar{p}_3(\lambda)$ corresponding to the primaries $[P_1(\lambda)], [P_2(\lambda)], [P_3(\lambda)]$. Then

$$\begin{aligned} C(\lambda) &= C_{\mathcal{K}'}(\lambda) + C_{\mathcal{K}}(\lambda) \\ &= C_1 P_1(\lambda) + C_2 P_2(\lambda) + C_3 P_3(\lambda) + C_{\mathcal{K}}(\lambda) \end{aligned} \tag{3.21}$$

where the C_i are given by Eq. (3.9) and

$$C_{\mathcal{K}}(\lambda) = C(\lambda) - (C_1 P_1(\lambda) + C_2 P_2(\lambda) + C_3 P_3(\lambda)). \tag{3.22}$$

Since $\mathcal{A}$ is infinite dimensional and $\mathcal{K}'$ has dimension 3, it follows that $\mathcal{K}$ is infinite dimensional, being three dimensions short of $\mathcal{A}$. Furthermore, $\mathcal{K}$ contains elements in both $\mathcal{A}_c$ and $\mathcal{A}_d$. In fact, we can write $\mathcal{K} = \mathcal{K}_c \oplus \mathcal{K}_d$ where $\mathcal{K}_c = \mathcal{K} \cap \mathcal{A}_c$ and $\mathcal{K}_d = \mathcal{K} \cap \mathcal{A}_d$. It is straightforward to find an arbitrary number of elements in $\mathcal{K}_d$. Choose any $K \geq 4$ wavelengths from $\mathcal{V}$ and let $K(\lambda) = \sum_{i=1}^{K} \alpha_i \delta(\lambda - \lambda_i)$. Assume an arbitrary basis for $\mathcal{C}$ with associated color matching functions $\bar{p}_i(\lambda), i = 1, 2, 3$, so that $K_j = \sum_{i=1}^{K} \alpha_i \bar{p}_j(\lambda_i), j = 1, 2, 3$. Then any solution of

$$\begin{bmatrix} \bar{p}_1(\lambda_1) & \cdots & \bar{p}_1(\lambda_K) \\ \bar{p}_2(\lambda_1) & \cdots & \bar{p}_2(\lambda_K) \\ \bar{p}_3(\lambda_1) & \cdots & \bar{p}_3(\lambda_K) \end{bmatrix} \begin{bmatrix} \alpha_1 \\ \vdots \\ \alpha_K \end{bmatrix} = \begin{bmatrix} 0 \\ 0 \\ 0 \end{bmatrix}$$

will give an element of $\mathcal{K}_d$. The null space of the matrix above is generally of dimension $K - 3$ (except in the unlikely case when the rank is less than 3) so there is an infinite number of solutions for each choice of $K \geq 4$ wavelengths. However, some of the α_i will always be negative.

The component $\mathcal{K}_c$ of the black space consists of solutions to the simultaneous linear integral equations

$$\int_{\mathcal{V}} K(\lambda) \bar{p}_i(\lambda) \, d\lambda = 0 \qquad i = 1, 2, 3, \tag{3.23}$$

where $K(\lambda) \in \mathcal{A}_c$. Since the $\bar{p}_i(\lambda)$ are empirically determined functions with no particular analytical representation, these equations must be discretized and solved numerically. Many researchers have sought suitable representations of black space and ways to find its members. Wyszecki and Stiles [61] (Section 3.8) give an overview of many approaches. Other notable contributions are due to Schmitt [44] and Cohen and Kappauf [4], among others. The approach presented here is somewhat more general than any of these.

Assume that

$$\bar{p}_i(\lambda) \approx \sum_{k=1}^{M} \bar{p}_{ik} \phi(\lambda - \lambda_k), \tag{3.24}$$

where $\{\lambda_k, k = 1, \ldots, M\} \subset \mathcal{V}$ is a set of wavelengths, typically uniformly spaced in $\mathcal{V}$, and $\phi(\lambda)$ is a local interpolation function. Since the $\bar{p}_i(\lambda)$ are typically very smooth and not known that precisely, this representation can be quite accurate for a sufficiently large M, and it may even be considered to be exact for certain standardized observers as will be seen later. Similarly, let

$$K(\lambda) = \sum_{l=1}^{N} K_l \psi(\lambda - \mu_l), \tag{3.25}$$

where $\{\mu_l, l = 1, \ldots, N\} \subset \mathcal{V}$ is another set of sample wavelengths. Note that $K(\lambda)$ and $\bar{p}_i(\lambda)$ belong to different spaces. Also, the $\bar{p}_i(\lambda)$ are fixed and known whereas the $K(\lambda)$ belong to the infinite dimensional space $\mathcal{K}_c$. Note the Eq. (3.25) is exact; however, N can be arbitrarily large, and it will be different from M, in general. Similarly, $\psi(\lambda) \in \mathcal{A}_c$ will generally be different from $\phi(\lambda)$.

Since $\psi(\lambda) \in \mathcal{A}_c$, all $K(\lambda)$ given by Eq. (3.25) will automatically be elements of $\mathcal{A}_c$. We denote

$$\mathcal{K}^{(N)} = \{K(\lambda) = \sum_{l=1}^{N} K_l \psi(\lambda - \mu_l) \mid K_l \in \mathbb{R}, \ K(\lambda) \in \mathcal{K}\} \tag{3.26}$$

as the N-dimensional approximation to $\mathcal{K}_c$, where elements of $\mathcal{K}^{(N)}$ are in one-to-one correspondence with elements of $\mathbb{R}^N$.

With the above representations, the integral equations of Eq. (3.23) can be written

$$\int_{\mathcal{V}} \left(\sum_{l=1}^{N} K_l \psi(\lambda - \mu_l) \right) \left(\sum_{k=1}^{M} \bar{p}_{ik} \phi(\lambda - \lambda_k) \right) d\lambda = 0$$
$$\sum_{l=1}^{N} \sum_{k=1}^{M} \bar{p}_{ik} \int_{\mathcal{V}} \psi(\lambda - \mu_l) \phi(\lambda - \lambda_k) \, d\lambda K_l = 0. \tag{3.27}$$

Since $\psi(\lambda)$ and $\phi(\lambda)$ are known analytical functions with small support, the $M \times N$ matrix Ψ with

$$\Psi_{kl} = \int_{\mathcal{V}} \psi(\lambda - \mu_l) \phi(\lambda - \lambda_k) \, d\lambda \tag{3.28}$$

can be analytically computed; because the support of ψ and ϕ are both small, most values will be zero. Defining the $3 \times M$ matrix $\bar{\mathbf{P}} = [\bar{p}_{ik}]$, we obtain $\bar{\mathbf{P}}\Psi\mathbf{K}^{(N)} = \mathbf{0}$. The solutions $\mathbf{K}^{(N)}$ will lie in the $N - 3$ dimensional null space of $\bar{\mathbf{P}}\Psi$, which can be found using standard methods employing the singular value decomposition [15]. Then, elements of $\mathcal{K}_c$ are obtained using Eq. (3.25). Note that if the expansion of the $\bar{p}_i(\lambda)$ is exact, then the resulting elements of $\mathcal{K}_c$ are exact. Using the above methods, we can generate an arbitrarily large number of continuous and discrete elements of black space.

3.7   CHANGE OF PRIMARIES

Most color imaging applications involve more than one set of primaries. For example, in additive display systems, there is a device-independent standardized basis (as will be introduced shortly) and a device-dependent basis specific to the display device (such as the CRT RGB phosphors we have been using as an example). We need to be able to convert the representation of any color with respect to one basis to its representation with respect to another basis. This is a *change-of-basis* operation in the context of linear algegra. Let $\mathcal{B} = \{[\mathbf{P}_1], [\mathbf{P}_2], [\mathbf{P}_3]\}$ be one basis for $\mathcal{C}$ and let $\tilde{\mathcal{B}} = \{[\tilde{\mathbf{P}}_1], [\tilde{\mathbf{P}}_2], [\tilde{\mathbf{P}}_3]\}$ be a different basis. An arbitrary color $[\mathbf{C}]$ has a unique representation with respect to each basis:

$$\begin{aligned}[\mathbf{C}] &= C_1[\mathbf{P}_1] + C_2[\mathbf{P}_2] + C_3[\mathbf{P}_3] \\ &= \tilde{C}_1[\tilde{\mathbf{P}}_1] + \tilde{C}_2[\tilde{\mathbf{P}}_2] + \tilde{C}_3[\tilde{\mathbf{P}}_3]. \end{aligned} \tag{3.29}$$

Typically, we know one set of coordinates (tristimulus values), and we want to find the other set.

We can express each of the primaries $[\mathbf{P}_j]$ in terms of the $[\tilde{\mathbf{P}}_k]$ as

$$[\mathbf{P}_j] = \sum_{k=1}^{3} a_{kj} [\tilde{\mathbf{P}}_k], \quad j = 1, 2, 3. \tag{3.30}$$

This allows us to transform the representations of arbitrary colors. Specifically,

$$
\begin{aligned}
[\mathbf{C}] &= \sum_{j=1}^{3} C_j [\mathbf{P}_j] \\
&= \sum_{j=1}^{3} C_j \sum_{k=1}^{3} a_{kj} [\tilde{\mathbf{P}}_k] \\
&= \sum_{k=1}^{3} \left(\sum_{j=1}^{3} C_j a_{kj} \right) [\tilde{\mathbf{P}}_k],
\end{aligned} \tag{3.31}
$$

from which we identify

$$\tilde{C}_k = \sum_{j=1}^{3} C_j a_{kj}, \quad k = 1, 2, 3. \tag{3.32}$$

This can be written in matrix form as

$$
\begin{bmatrix} \tilde{C}_1 \\ \tilde{C}_2 \\ \tilde{C}_3 \end{bmatrix} = \begin{bmatrix} a_{11} & a_{12} & a_{13} \\ a_{21} & a_{22} & a_{23} \\ a_{31} & a_{32} & a_{33} \end{bmatrix} \begin{bmatrix} C_1 \\ C_2 \\ C_3 \end{bmatrix} \tag{3.33}
$$

or $\mathbf{C}_{\tilde{\mathcal{B}}} = \mathbf{A}\mathbf{C}_{\mathcal{B}}$. Furthermore, it is clear that $\mathbf{C}_{\mathcal{B}} = \mathbf{A}^{-1}\mathbf{C}_{\tilde{\mathcal{B}}}$.

Since we can carry out such a change of basis between any two valid bases, the matrix should identify the bases. There is no standard notation for this, and just about any choice is awkward. I will adopt the notation $\mathbf{C}_{\tilde{\mathcal{B}}} = \mathbf{A}_{\mathcal{B} \to \tilde{\mathcal{B}}} \mathbf{C}_{\mathcal{B}}$ to identify the matrix to transform tristimulus values without ambiguity. Using this notation, we see that $\mathbf{A}_{\tilde{\mathcal{B}} \to \mathcal{B}} = (\mathbf{A}_{\mathcal{B} \to \tilde{\mathcal{B}}})^{-1}$.

This matrix operation can also be used to transform the color matching functions for the primaries $[\mathbf{P}_j]$ to color matching functions for the primaries $[\tilde{\mathbf{P}}_k]$, recognizing that color matching functions specify tristimulus values for each λ:

$$
\begin{bmatrix} \bar{\tilde{p}}_1(\lambda) \\ \bar{\tilde{p}}_2(\lambda) \\ \bar{\tilde{p}}_3(\lambda) \end{bmatrix} = \mathbf{A}_{\mathcal{B} \to \tilde{\mathcal{B}}} \begin{bmatrix} \bar{p}_1(\lambda) \\ \bar{p}_2(\lambda) \\ \bar{p}_3(\lambda) \end{bmatrix}. \tag{3.34}
$$

This can be viewed as an ordinary matrix equation for real numbers evaluated separately at each wavelength λ. However, it can just as well be viewed as a matrix multiplication of a real matrix by a column matrix of elements from the vector space of real functions of λ on $\mathcal{V}$.

The relationship between primaries in Eq. (3.30) can also be written in matrix form as

$$
\begin{bmatrix} [\mathbf{P}_1] \\ [\mathbf{P}_2] \\ [\mathbf{P}_3] \end{bmatrix} = \mathbf{A}^T_{\mathcal{B} \to \tilde{\mathcal{B}}} \begin{bmatrix} [\tilde{\mathbf{P}}_1] \\ [\tilde{\mathbf{P}}_2] \\ [\tilde{\mathbf{P}}_3] \end{bmatrix}. \tag{3.35}
$$

Again, this is not a conventional real matrix multiplication, but the multiplication of a real 3×3 matrix with a 3×1 column matrix of elements of the vector space $\mathcal{C}$. This is perfectly well defined.

There are six transformations just defined, counting both directions, and they are often confused. Table 3.1 summarizes these six transformations, all defined in terms of the matrix $\mathbf{A}_{\mathcal{B} \to \tilde{\mathcal{B}}}$.

Table 3.1: Conversions of tristimulus values, color matching functions and primaries between two sets of primaries defined by bases $\mathcal{B}$ and $\tilde{\mathcal{B}}$.

1.	$\begin{bmatrix} \tilde{C}_1 \\ \tilde{C}_2 \\ \tilde{C}_3 \end{bmatrix} = \mathbf{A}_{\mathcal{B} \to \tilde{\mathcal{B}}} \begin{bmatrix} C_1 \\ C_2 \\ C_3 \end{bmatrix}$	4.	$\begin{bmatrix} C_1 \\ C_2 \\ C_3 \end{bmatrix} = \mathbf{A}^{-1}_{\mathcal{B} \to \tilde{\mathcal{B}}} \begin{bmatrix} \tilde{C}_1 \\ \tilde{C}_2 \\ \tilde{C}_3 \end{bmatrix}$
2.	$\begin{bmatrix} \bar{\tilde{p}}_1(\lambda) \\ \bar{\tilde{p}}_2(\lambda) \\ \bar{\tilde{p}}_3(\lambda) \end{bmatrix} = \mathbf{A}_{\mathcal{B} \to \tilde{\mathcal{B}}} \begin{bmatrix} \bar{p}_1(\lambda) \\ \bar{p}_2(\lambda) \\ \bar{p}_3(\lambda) \end{bmatrix}$	5.	$\begin{bmatrix} \bar{p}_1(\lambda) \\ \bar{p}_2(\lambda) \\ \bar{p}_3(\lambda) \end{bmatrix} = \mathbf{A}^{-1}_{\mathcal{B} \to \tilde{\mathcal{B}}} \begin{bmatrix} \bar{\tilde{p}}_1(\lambda) \\ \bar{\tilde{p}}_2(\lambda) \\ \bar{\tilde{p}}_3(\lambda) \end{bmatrix}$
3.	$\begin{bmatrix} [\tilde{\mathbf{P}}_1] \\ [\tilde{\mathbf{P}}_2] \\ [\tilde{\mathbf{P}}_3] \end{bmatrix} = \mathbf{A}^{-T}_{\mathcal{B} \to \tilde{\mathcal{B}}} \begin{bmatrix} [\mathbf{P}_1] \\ [\mathbf{P}_2] \\ [\mathbf{P}_3] \end{bmatrix}$	6.	$\begin{bmatrix} [\mathbf{P}_1] \\ [\mathbf{P}_2] \\ [\mathbf{P}_3] \end{bmatrix} = \mathbf{A}^{T}_{\mathcal{B} \to \tilde{\mathcal{B}}} \begin{bmatrix} [\tilde{\mathbf{P}}_1] \\ [\tilde{\mathbf{P}}_2] \\ [\tilde{\mathbf{P}}_3] \end{bmatrix}$

3.8 THE VISUAL SUBSPACE AND GENERAL COLOR SPACES

The functions $C_i = \int_{\mathcal{V}} C(\lambda) \bar{p}_i(\lambda) \, d\lambda$ are *linear functionals* on $\mathcal{A}$, i.e., linear maps from $\mathcal{A}$ to $\mathbb{R}$. The set of all linear functionals on $\mathcal{A}$ is a vector space $\mathcal{A}^*$ called the dual space of $\mathcal{A}$ [19]. We have seen that normal human vision is associated with three such linear functionals specified by the color matching functions $\bar{p}_i(\lambda)$, $i = 1, 2, 3$, for a specific basis $\mathcal{B} = \{[\mathbf{P}_1], [\mathbf{P}_2], [\mathbf{P}_3]\}$. For any other basis, the new color matching functions are a linear combination of the $\bar{p}_i(\lambda)$ according to Eq. (3.34). Thus all sets of color matching functions lie in a three-dimensional space spanned by the $\bar{p}_i(\lambda)$.

Although the set of all linear functionals on $\mathcal{A}$ is much larger, we will consider only those of the form $\int_{\mathcal{V}} C(\lambda) f(\lambda) \, d\lambda$ where $f(\lambda)$ is an element of the space of continuous functions on $\mathcal{V}$, denoted $\mathcal{C}(\mathcal{V})$. Color matching functions are usually very smooth, such as those shown in Fig. 3.4. We do not consider the possibility of color matching functions that are discontinuous or that involve Dirac deltas. Thus, we can associate each linear functional with an element of the set of continuous functions on $\mathcal{V}$. This identifies a subspace of $\mathcal{A}^*$ isomorphic to $\mathcal{C}(\mathcal{V})$ that we call $\mathcal{A}^*_{\mathcal{C}}$. We thus informally refer to the color matching functions as elements of $\mathcal{A}^*_{\mathcal{C}}$.

The subspace of $\mathcal{A}^*_{\mathcal{C}}$ spanned by $\{\bar{p}_i(\lambda), i = 1, 2, 3\}$ *characterizes* the color space, and different choices of the subspace define different color spaces, which could correspond to different individuals. By allowing dimensions other than 3, we can describe the color spaces of individuals with color deficiencies, different creatures (maybe), or electronic cameras. We denote this subspace $\mathcal{VS}_{\mathcal{C}}$ for *Visual Subspace* corresponding to the color space $\mathcal{C}$

$$\mathcal{VS}_{\mathcal{C}} = \mathrm{span}(\bar{p}_1(\lambda), \bar{p}_2(\lambda), \bar{p}_3(\lambda)) \subset \mathcal{A}^*_{\mathcal{C}}. \tag{3.36}$$

We now show how a color space can be defined starting from it visual subspace, namely an arbitrary finite dimensional subspace of $\mathcal{C}(\mathcal{V})$. This is a common way to introduce the theory of color spaces although it is often based on a finite-dimensional stimulus space (rather than $\mathcal{A}$) and uses matrix theory [46]. To distinguish from the human trichromatic color space discussed so far, we will call the color space to be developed $\mathcal{D}$ (which has no particular signification).

Let $\mathcal{VS}_{\mathcal{D}}$ be a finite-dimensional subspace of $\mathcal{A}^*_{\mathcal{C}}$; specifically, let $\dim(\mathcal{VS}_{\mathcal{D}}) = M$ and let $\{\bar{p}_1(\lambda), \ldots, \bar{p}_M(\lambda)\}$ be a basis for $\mathcal{VS}_{\mathcal{D}}$. Generally, $\mathcal{VS}_{\mathcal{D}}$ will be *specified* by such a basis. For $C_1(\lambda), C_2(\lambda) \in \mathcal{A}$, we say that

$$
\begin{aligned}
&C_1(\lambda) \boxminus C_2(\lambda) \quad \text{if and only if} \\
&\int_{\mathcal{V}} C_1(\lambda) f(\lambda)\, d\lambda = \int_{\mathcal{V}} C_2(\lambda) f(\lambda)\, d\lambda \quad \text{for all } f(\lambda) \in \mathcal{VS}_{\mathcal{D}}.
\end{aligned}
\tag{3.37}
$$

This holds if and only if

$$\int_{\mathcal{V}} C_1(\lambda) \bar{p}_i(\lambda)\, d\lambda = \int_{\mathcal{V}} C_2(\lambda) \bar{p}_i(\lambda)\, d\lambda, \quad i = 1, \ldots, M. \tag{3.38}$$

The relation $\boxminus$ is easily seen to be an equivalence relation on $\mathcal{A}$. The set $[C(\lambda)]_{\boxminus}$ is the equivalence class containing $C(\lambda)$, and the set of all such equivalence classes forms a partition of $\mathcal{A}$. From the definition, it is clear that

$$
\begin{aligned}
C_1(\lambda) \boxminus C_2(\lambda) &\Rightarrow \alpha C_1(\lambda) \boxminus \alpha C_2(\lambda) \quad \text{for all } \alpha \in \mathbb{R}; \\
C(\lambda) + C_1(\lambda) &\boxminus C(\lambda) + C_2(\lambda) \quad \text{if and only if } C_1(\lambda) \boxminus C_2(\lambda).
\end{aligned}
$$

Thus, equivalence classes can be scaled and added, and it is easily seen that all axioms of a vector space are satisfied. Thus

$$\mathcal{D} = \{[C(\lambda)]_{\boxminus} \mid C(\lambda) \in \mathcal{A}\}, \tag{3.39}$$

where we continue to use the notation [**C**] to denote an element of $\mathcal{D}$.

Theorem 3.11   *The vector space $\mathcal{D}$ generated by $\mathcal{VS}_\mathcal{D} = \mathrm{span}(\bar{p}_1(\lambda), \ldots, \bar{p}_M(\lambda))$ is of dimension M. There is a unique basis $\mathcal{B} = [\mathbf{P}_1], \ldots, [\mathbf{P}_M]$ for which $\{\bar{p}_1(\lambda), \ldots, \bar{p}_M(\lambda)\}$ is the corresponding set of color matching functions.*

Proof. There is a well-defined mapping from $\mathcal{D}$ to $\mathbb{R}^M$ given by

$$\mathcal{S}_{\mathcal{D}\to\mathbb{R}^M} : [C(\lambda)]_\boxminus \mapsto \begin{bmatrix} \int_\mathcal{V} C(\lambda)\bar{p}_1(\lambda)\,d\lambda \\ \vdots \\ \int_\mathcal{V} C(\lambda)\bar{p}_M(\lambda)\,d\lambda \end{bmatrix} \tag{3.40}$$

which by definition is independent of which $C(\lambda)$ in the equivalence class is used. This is clearly a linear map. It is *one-to-one* (monic) since, by definition,

$$\mathcal{S}_{\mathcal{D}\to\mathbb{R}^M}([C_1(\lambda)]_\boxminus) = \mathcal{S}_{\mathcal{D}\to\mathbb{R}^M}([C_2(\lambda)]_\boxminus) \quad \text{implies} \quad [C_1(\lambda)]_\boxminus = [C_2(\lambda)]_\boxminus.$$

It is *onto* (epic) since for any $1 \le i \le M$, we can find a $P_i(\lambda) \ne 0$ such that $\int_\mathcal{V} P_i(\lambda)\bar{p}_k(\lambda)\,d\lambda = 0$ for all $k \ne i$. This could be done by finding suitable elements of $\mathcal{A}_d$ as was described for finding elements of the black space in Section 3.6 and is guaranteed to be possible since the $\bar{p}_k(\lambda)$ are linearly independent. By suitable scaling, we can also ensure that $\int_\mathcal{V} P_i(\lambda)\bar{p}_i(\lambda)\,d\lambda = 1$. It follows that for any $\mathbf{a} = \begin{bmatrix} a_1 & \cdots & a_M \end{bmatrix}^T \in \mathbb{R}^M$, $\mathcal{S}_{\mathcal{D}\to\mathbb{R}^M}(a_1 P_1(\lambda) + \cdots + a_M P_M(\lambda)) = \mathbf{a}$. Thus $\mathcal{S}_{\mathcal{D}\to\mathbb{R}^M}$ is an isomorphism and so $\dim(\mathcal{D}) = M$. Furthermore, $[\mathbf{P}_i] = [P_i(\lambda)]_\boxminus$, $i = 1, \ldots, M$ forms the desired reciprocal basis. $\qquad\square$

The mapping $\mathcal{S} : \mathcal{A} \to \mathcal{D} : C(\lambda) \mapsto [C(\lambda)]_\boxminus$ is a vector space homomorphism. $\mathcal{K} = \ker \mathcal{S} \subset \mathcal{A}$ is the set of all $C(\lambda) \in \mathcal{A}$ mapping to $\mathbf{0} \in \mathcal{D}$ and is called the black subspace for the given color space. We have that $C_1(\lambda) \boxminus C_2(\lambda)$ if and only if $C_1(\lambda) - C_2(\lambda) \in \mathcal{K}$. $\mathcal{D}$ is isomorphic to the factor space $\mathcal{A}/\mathcal{K}$, and we can decompose $\mathcal{A}$ as $\mathcal{A} = \mathcal{K} \oplus \mathcal{K}'$ where $\mathcal{K}'$ is of dimension M and is isomorphic to $\mathcal{D}$; however, $\mathcal{K}'$ is *not* unique. Everything described previously on change of basis applies in exactly the same way but with 3 replaced by M.

We conclude with a result on embedded color spaces.

Theorem 3.12   *Suppose that $\mathcal{D}_1$ and $\mathcal{D}_2$ are two color spaces corresponding to $\mathcal{VS}_1$ and $\mathcal{VS}_2$ of dimension M_1 and M_2, respectively, where $M_2 < M_1$. Assume that $C_1(\lambda) \boxminus_1 C_2(\lambda) \Rightarrow C_1(\lambda) \boxminus_2 C_2(\lambda)$. Then $\mathcal{VS}_2 \subset \mathcal{VS}_1$.*

Proof. The statement that color matches in $\mathcal{D}_1$ are also matches in $\mathcal{D}_2$ means $[C(\lambda)]_{\boxminus_1} \subset [C(\lambda)]_{\boxminus_2}$. Thus, there is a well-defined map $\mathcal{S}_{\mathcal{D}_1\to\mathcal{D}_2} : [C(\lambda)]_{\boxminus_1} \mapsto [C(\lambda)]_{\boxminus_2}$. This is easily seen to be a linear map. Let $\mathcal{A} = \{[\mathbf{Q}_1], \ldots, [\mathbf{Q}_{M_1}]\}$ be a basis for $\mathcal{D}_1$ and $\mathcal{B} = \{[\mathbf{P}_1], \ldots, [\mathbf{P}_{M_2}]\}$ be a basis for $\mathcal{D}_2$. Then the map $\mathcal{S}_{\mathcal{D}_1 \to \mathcal{D}_2}$ has an $M_2 \times M_1$ matrix $\mathbf{A}_{\mathcal{A} \to \mathcal{B}}$ such that $\mathbf{C}_{\mathcal{B}} = \mathbf{A}_{\mathcal{A} \to \mathcal{B}} \mathbf{C}_{\mathcal{A}}$. It follows that

$$\begin{bmatrix} \bar{p}_1(\lambda) \\ \vdots \\ \bar{p}_{M_2}(\lambda) \end{bmatrix} = \mathbf{A}_{\mathcal{A} \to \mathcal{B}} \begin{bmatrix} \bar{q}_1(\lambda) \\ \vdots \\ \bar{q}_{M_1}(\lambda) \end{bmatrix}. \tag{3.41}$$

Thus, each of the basis vectors of $\mathcal{VS}_2$ is a linear combination of basis vectors of $\mathcal{VS}_1$, and so $\mathcal{VS}_2 \subset \mathcal{VS}_1$. □

The expressions $C_i = \int_{\mathcal{V}} C(\lambda) \bar{p}_i(\lambda)\, d\lambda$ are often considered to be inner products on the space of functions of λ. I do not believe that this is a correct interpretation. The functions $C(\lambda)$ and $\bar{p}(\lambda)$ are different types of objects. The $C(\lambda)$ are spectral densities belonging to $\mathcal{A}$, which may include Dirac deltas and not be square integrable. The color matching functions are weighting functions, and, typically, they can be considered to be very smooth. Thus I will not interpret the computation of tristimulus values as an inner product on $\mathcal{A}$ but as a linear functional as described above. (Note that this is sometimes called an inner product between elements of $\mathcal{A}$ and elements of $\mathcal{A}^*$.) Krantz [26] and Koenderink and van Doorn [25] have clearly commented on this interpretation.

3.9   THE CIE COLOR SPACES

As we have seen, a color space is characterized by the subspace of $\mathcal{A}^*$ spanned by three color matching functions. These color matching functions can be determined by experiment with human observers for any convenient basis and, subsequently, converted to correspond to any desired basis using Eq. (3.34). Over the years, numerous experiments have been conducted to determine color matching functions for different sets of primaries and with different experimental conditions. Each result corresponds to a different and *bona fide* color space as we have defined it. In 1931, the Commission Internationale d'Éclairage (CIE) standardized a particular color space called the *CIE 1931 standard colorimetric observer*. The color matching functions were provided for two sets of primaries: a set of monochromatic red, green and blue primaries, and a transformed basis referred to as XYZ. The latter basis is the most widely used standard basis for colorimetry.

The RGB basis will be referred to here as $\mathcal{RGB}31$:

$$\mathcal{RGB}31 = \{k_R[\delta(\lambda - 700.0)], k_G[\delta(\lambda - 546.1)], k_B[\delta(\lambda - 435.8)]\}. \tag{3.42}$$

The constants k_R, k_G, k_B were chosen so that the sum of one unit of each of these primaries would match an equal energy white $[\mathbf{E}] = k_E[E(\lambda)]$, where $E(\lambda) = 1$. The color matching functions were based on experimental results by W.D. Wright using ten observers and J. Guild using seven observers. Wright provides an interesting account of the standardization of this basis by the CIE [60]. The standardized color matching functions are depicted in Fig. 3.5; they are similar in form (but different in detail) to the RGB CRT phosphors color matching functions of Fig. 3.4.

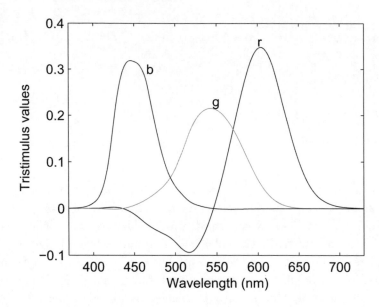

Figure 3.5: Color matching functions of the CIE 1931 RGB primaries, monochromatic lights at wavelengths 700.0 nm, 546.1 nm and 435.8 nm, respectively.

For various practical reasons designed to simplify the computation of tristimulus values with 1930s technology, an alternate set of primaries called XYZ was introduced. In particular, the XYZ color matching functions were positive everywhere, and the Y color matching function was equal to the relative luminous efficiency curve $V(\lambda)$ of photometry (this will be introduced later in this lecture). The resulting primaries are not physically realizable colors. We denote this basis

$$\mathcal{XYZ} = \{[\mathbf{X}], [\mathbf{Y}], [\mathbf{Z}]\}. \tag{3.43}$$

The change of basis matrix to convert tristimulus values from the $\mathcal{RGB}31$ basis to the $\mathcal{XYZ}$ basis is [43]

$$\mathbf{A}_{\mathcal{RGB}31\to\mathcal{XYZ}} = \begin{bmatrix} 2.768892 & 1.751748 & 1.130160 \\ 1.000000 & 4.590700 & 0.060100 \\ 0.000000 & 0.056508 & 5.594292 \end{bmatrix}. \tag{3.44}$$

We can of course use this matrix and its various derivatives to transform tristimulus values, primaries and color matching functions in both directions. However, the $\mathcal{RGB}31$ basis is rarely used per se; all other bases are defined in terms of $\mathcal{XYZ}$. The XYZ color matching functions are obtained by

$$\begin{bmatrix} \bar{x}(\lambda) \\ \bar{y}(\lambda) \\ \bar{z}(\lambda) \end{bmatrix} = \mathbf{A}_{\mathcal{RGB}31\to\mathcal{XYZ}} \begin{bmatrix} \bar{r}_{31}(\lambda) \\ \bar{g}_{31}(\lambda) \\ \bar{b}_{31}(\lambda) \end{bmatrix} \tag{3.45}$$

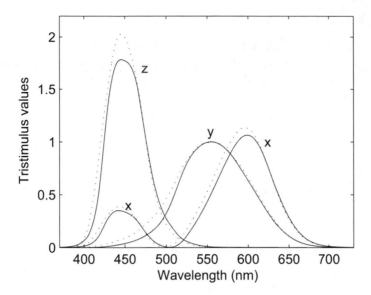

Figure 3.6: Color matching functions of the CIE 1931 (solid) and 1964(dotted) XYZ primaries.

and are shown in Fig. 3.6.

The CIE 1931 XYZ color matching functions have been standardized. They are given for values of λ ranging from 360nm to 830nm at 1nm increments and with seven digits. They are assumed to be linearly interpolated within the 1nm intervals, and thus they can be considered as defined for all wavelengths $\lambda \in V$. This high precision does not signify this level of accuracy but is provided to allow repeatability. The data are widely available and can be found in the standard treatise by Wyszecki and Stiles [61] in Table I(3.3.1) and online at the website maintained by the Color & Vision Research Laboratories, UCL Institute of Ophthamology, London, England [51]. I have used the data from this website to draw the figures in this work.

The CIE 1931 standard observer is meant to apply to small stimuli in the range $1° - 4°$ of visual angle, and it is sometimes referred to as the $2°$ standard colorimetric observer. To cover the slightly different case of larger stimuli, the CIE introduced the $10°$ colorimetric observer in 1964. Once again, color matching functions are standardized and available in the same sources mentioned above. Both RGB and XYZ bases and color matching functions were introduced. To distinguish with the 1931 XYZ basis, the $10°$ basis vectors are referred to as

$$\mathcal{XYZ}_{10} = \{[\mathbf{X}_{10}], [\mathbf{Y}_{10}], [\mathbf{Z}_{10}]. \tag{3.46}$$

The $\mathcal{XYZ}_{10}$ color matching functions are also shown in Fig. 3.6 (dotted lines). An important point to make here is that these two systems represent *different* color spaces. The two sets of color matching functions span different subspaces of the dual space $\mathcal{A}^*$. I will come back to transformations between different color spaces later.

3.10   PHYSICALLY REALIZABLE COLORS

3.10.1   THE CONE OF PHYSICALLY REALIZABLE COLORS

An element $[\mathbf{C}]$ of $\mathcal{C}$ is said to be a physically realizable color, or simply a real color, if at least one element of the equivalence class $[\mathbf{C}]$ belongs to $\mathcal{P}$. Thus we can define the set of all physically realizable colors to be

$$\mathcal{C}_R = \{[C(\lambda)] \mid C(\lambda) \in \mathcal{P}\}, \tag{3.47}$$

or more simply $\mathcal{C}_R = \mathcal{S}(\mathcal{P})$, the image of $\mathcal{P}$ under $\mathcal{S}$.

Theorem 3.13   *The set of physically realizable colors $\mathcal{C}_R$ is a convex subset of $\mathcal{C}$ that has the structure of a convex cone.*

Proof. Let $[\mathbf{C}_1], [\mathbf{C}_2] \in \mathcal{C}_R$. By the definition of a convex set, we need to show that $\alpha[\mathbf{C}_1] + (1-\alpha)[\mathbf{C}_2] \in \mathcal{C}_R$ for all $0 \le \alpha \le 1$. By hypothesis, there exist $C_1(\lambda), C_2(\lambda) \in \mathcal{P}$ such that $[\mathbf{C}_1] = [C_1(\lambda)]$ and $[\mathbf{C}_2] = [C_2(\lambda)]$. It is clear from the definition of $\mathcal{P}$ that $\alpha C_1(\lambda) + (1-\alpha)C_2(\lambda) \in \mathcal{P}$ for any $\alpha \in [0, 1]$, and thus $[\alpha C_1(\lambda) + (1-\alpha)C_2(\lambda)] \in \mathcal{C}_R$. Since $[\alpha C_1(\lambda) + (1-\alpha)C_2(\lambda)] = \alpha[C_1(\lambda)] + (1-\alpha)[C_2(\lambda)] = \alpha[\mathbf{C}_1] + (1-\alpha)[\mathbf{C}_2]$, we conclude that $\mathcal{C}_R$ is a convex set. It is a convex cone since $C(\lambda) \in \mathcal{P}$ implies that $\alpha C(\lambda) \in \mathcal{P}$ for $\alpha \ge 0$ (i.e., $\mathcal{P}$ is a convex cone) and thus $[\mathbf{C}] \in \mathcal{C}_R$ implies $\alpha[\mathbf{C}] \in \mathcal{C}_R$ for $\alpha \ge 0$. $\square$

The cone $\mathcal{C}_R$ can be characterized as the convex hull of the set of monochromatic lights $\mathcal{M} = \{\alpha[\delta(\lambda - \mu)] \mid \mu \in \mathcal{V}, \alpha \in \mathbb{R}_+\}$. The convex hull of a set of points in a vector space is the smallest convex set that contains all the points. It is the intersection of all convex sets that contain these points. We use the following specific property of a convex cone.

Proposition 3.14   *If Q is a convex cone and $q_1, q_2, \ldots, q_K \in Q$, then $\alpha_1 q_1 + \alpha_2 q_2 + \cdots + \alpha_K q_K \in Q$ if $\alpha_k \ge 0, k = 1, 2, \ldots, K$.*

Proof. First we show that if $r_1, r_2 \in Q$, then $r_1 + \alpha r_2 \in Q$ if $\alpha \ge 0$. By convexity of Q,

$$\frac{1}{1+\alpha}r_1 + \frac{\alpha}{1+\alpha}r_2 \in Q, \tag{3.48}$$

and multiplying by $1 + \alpha$, $r_1 + \alpha r_2 \in Q$ since Q is a convex cone. The proof follows simply by induction: $q_1 \in Q \Rightarrow \alpha_1 q_1 \in Q \Rightarrow \alpha_1 q_1 + \alpha_2 q_2 \in Q \Rightarrow \alpha_1 q_1 + \alpha_2 q_2 + \alpha_3 q_3 \in Q$, etc. $\square$

Theorem 3.15   *The set of physically realizable colors $\mathcal{C}_R$ is the convex hull of the set $\mathcal{M}$ of monochromatic lights.*

Proof. Let $\mathcal{H}$ denote the convex hull of $\mathcal{M}$. Since every element of $\mathcal{M}$ belongs to $\mathcal{C}_R$, and $\mathcal{C}_R$ is convex, it follows that $\mathcal{C}_R$ is a convex set containing $\mathcal{M}$ and so $\mathcal{H} \subset \mathcal{C}_R$. We now must show that any element $[\mathbf{C}]$ of $\mathcal{C}_R$ belongs to $\mathcal{H}$. Let $[\mathbf{C}] = [C(\lambda)]$ for some $C(\lambda) \in \mathcal{P}$. We can write $C(\lambda) = C_c(\lambda) + C_d(\lambda)$ where $C_c(\lambda)$ is a non-negative piecewise-continuous function and $C_d(\lambda) = \sum_{k=1}^{K} \alpha_k \delta(\lambda - \lambda_k)$, where $\lambda_k \in \mathcal{V}$ and $\alpha_k > 0$, $k = 1, 2, \ldots, K$. From Proposition 3.14, $[C_d(\lambda)] \in \mathcal{H}$. We can express $[\mathbf{C}_c]$ in terms of the monochromatic lights as

$$[\mathbf{C}_c] = \int_{\mathcal{V}} C_c(\mu)[\delta(\lambda - \mu)] \, d\mu. \tag{3.49}$$

Since $C_c(\mu)$ is non-negative and piecewise continuous, we can extend Proposition 3.14 to this case by continuity arguments. Thus, since every $[\delta(\lambda - \mu)]$ belongs to $\mathcal{H}$, $[\mathbf{C}_c]$ must belong to $\mathcal{H}$ and thus $[\mathbf{C}]$ belongs to $\mathcal{H}$.

$\square$

The set $\mathcal{C}_R$ defined above is isomorphic to the set $\mathcal{C}_P$ defined in Section 3.2. To any $[C(\lambda)]_{\triangle} \in \mathcal{C}_P$ we associate $[C(\lambda)]_{\boxminus} \in \mathcal{C}_R$, where $C(\lambda) \in \mathcal{P}$, and vice versa. The operations of addition and scalar multiplication are preserved in this assignment, making it an isomorphism. However, the equivalence classes comprising the two sets are not the same since $[C(\lambda)]_{\boxminus}$ contains elements of $\mathcal{A}$ that are not in $\mathcal{P}$. Thus, $[C(\lambda)]_{\triangle} \subset [C(\lambda)]_{\boxminus}$, and, in fact, $[C(\lambda)]_{\boxminus} = [C(\lambda)]_{\triangle} + \mathcal{K}$.

While the above properties apply to any color space, in any number of dimensions, the cone $\mathcal{C}_R$ for the trichromatic human visual color space as exemplified by the CIE 1931 standard observer has the following additional properties. We first need the following result, which will also be important later.

Theorem 3.16 *In the trichromatic visual color space, the only physically realizable element of black space is $K(\lambda) = 0$, i.e., $\mathcal{P} \cap \mathcal{K} = 0$.*

Proof. If $K(\lambda) \in \mathcal{K}$, it follows that $\int_{\mathcal{V}} K(\lambda) \bar{p}_i(\lambda) \, d\lambda = 0$ for $i = 1, 2, 3$, for any set of color matching functions for the human visual color space. In particular, this holds for $\{\bar{p}_i(\lambda)\} = \{\bar{x}(\lambda), \bar{y}(\lambda), \bar{z}(\lambda)\}$ as shown in Fig. 3.6. From the figure, it is clear that over any subinterval of $\mathcal{V}$, at least two of the color matching functions are strictly positive and all are non-negative. Thus, if $K(\lambda) \in \mathcal{P}$, $\int_{\mathcal{V}} K(\lambda) \bar{p}_i(\lambda) \, d\lambda \geq 0$, for $i = 1, 2, 3$, and they can all be simultaneously zero only if $K(\lambda) = 0$ for all $\lambda \in \mathcal{V}$. $\square$

We can now establish the following properties of $\mathcal{C}_R$ for the trichromatic human color space.

Theorem 3.17 *For the trichromatic human color space,*
(i) $\mathcal{C}_R$ lies on one side of a plane in $\mathcal{C}$;
(ii) The monochromatic lights $\mathcal{M}$ lie on the boundary of $\mathcal{C}_R$.

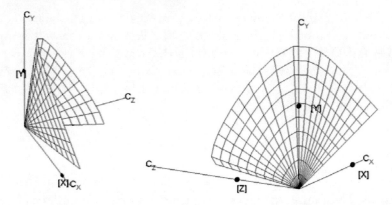

Figure 3.7: Two views illustrating the cone of physically realizable colors in the XYZ color space. It is the convex hull of the set of monochromatic lights that lie on its shark-fin shaped boundary, closed off by the plane of purples (at the bottom in the left view).

Proof. (i) Suppose that non-zero $[\mathbf{C}] \in \mathcal{C}_R$. Then, there exists $C(\lambda) \in \mathcal{P}$ such that $\mathcal{S}(C(\lambda)) = [\mathbf{C}]$ (or equivalently, there exists $C(\lambda) \in \mathcal{S}^{-1}([\mathbf{C}] \cap \mathcal{P})$. Suppose that $-[\mathbf{C}] \in \mathcal{C}_R$. Since $-C(\lambda) \notin \mathcal{P}$, there must be some non-zero $K(\lambda) \in \mathcal{K}$ such that $C_1(\lambda) = C(\lambda) + K(\lambda) \in \mathcal{P}$. Thus $K(\lambda) = -C(\lambda) + C_1(\lambda) \in \mathcal{P}$ and so $K(\lambda) = 0$ by Theorem 3.16. Thus, if $[\mathbf{C}] \in \mathcal{C}_R$, then $-[\mathbf{C}] \notin \mathcal{C}_R$, and there must be a plane through the origin so that $[\mathbf{C}] \in \mathcal{C}_R$ lies entirely on one side of it.

(ii) The statement that all monochromatic lights lie on the boundary of $\mathcal{C}_R$ is, in fact, an empirical result of the form of the color matching functions for human vision. Fig. 3.7 illustrates two views of the tristimulus values of the set of monochromatic lights, on the axes determined by the CIE1931 XYZ primaries. They all lie on the illustrated curved surface, and none are in the interior of the cone. □

The cone $\mathcal{C}_R$ consists of the region bounded by the curved surface and the plane of purples (not shown), which would close it off at the bottom (in the left view). The illustrated cone is arbitrarily extended out to the plane $C_X + C_Y + C_Z = 2$. The primaries [X], [Y] and [Z] are shown and are clearly outside $\mathcal{C}_R$ as they must be since they are not physically realizable. Schrödinger [45] showed an illustration of the cone $\mathcal{C}_R$ in an arbitrary set of coordinates, with the plane of purples as the base. Other views can be seen in [25].

3.10.2 ADDITIVE REPRODUCTION OF COLORS

As we have mentioned, the tristimulus values allow us to *numerically* specify colors in terms of a set of primaries, which may or may not be physical colors. However, to actually *synthesize* colors, say on a CRT, we add (superpose) lights corresponding to three physical primaries, weighted by the tristimulus values, which must be non-negative. Not all colors can be synthesized in this way, and the judicious choice of these *display primaries* determines what fraction of the set of all physical colors can in fact be synthesized with the given primaries. The following theorem identifies the set of colors that can be synthesized by a trichromatic additive display.

Theorem 3.18   *Let* $[\mathbf{P}_1]$, $[\mathbf{P}_2]$ *and* $[\mathbf{P}_3]$ *be three physically realizable and linearly independent colors. The set of all colors formed as linear combinations of* $[\mathbf{P}_1]$, $[\mathbf{P}_2]$ *and* $[\mathbf{P}_3]$ *with non-negative coefficients is a convex cone with a triangular cross section. It is the convex hull of the three half lines* $\alpha_1[\mathbf{P}_1]$, $\alpha_2[\mathbf{P}_2]$ *and* $\alpha_3[\mathbf{P}_3]$ *with* $\alpha_i \geq 0$, *for* $i = 1, 2, 3$.

Proof. Since $[\mathbf{P}_1]$, $[\mathbf{P}_2]$ and $[\mathbf{P}_3]$ are linearly independent, they form a basis of $\mathcal{C}$. Let

$$\mathcal{Q}_\mathcal{B} = \{\alpha_1[\mathbf{P}_1] + \alpha_2[\mathbf{P}_2] + \alpha_3[\mathbf{P}_3] \mid \alpha_i \geq 0, i = 1, 2, 3\}. \tag{3.50}$$

If $\mathcal{Q}$ is any convex cone that contains $[\mathbf{P}_1]$, $[\mathbf{P}_2]$ and $[\mathbf{P}_3]$, then by Proposition 3.14 $\mathcal{Q}_\mathcal{B} \subset \mathcal{Q}$, and thus $\mathcal{Q}_\mathcal{B}$ is a subset of the said convex hull. On the other hand, it is evident from the definitions that $\mathcal{Q}_\mathcal{B}$ is a convex cone containing the given lines, and so it must contain the convex hull. Thus, $\mathcal{Q}_\mathcal{B}$ is the convex hull of the three half lines. This cone is bounded by the three planes containing the pairs of lines $\{\alpha_1[\mathbf{P}_1], \alpha_2[\mathbf{P}_2]\}$, $\{\alpha_1[\mathbf{P}_1], \alpha_3[\mathbf{P}_3]\}$ and $\{\alpha_2[\mathbf{P}_2], \alpha_3[\mathbf{P}_3]\}$, respectively, and thus it has a triangular cross section.   □

Fig. 3.8(a) illustrates this triangular cone in the XYZ space for the CRT RGB primaries with color matching functions given in Fig. 3.4. Of course, it is entirely within the cone $\mathcal{C}_R$. The goal in choosing additive display primaries is to fill as much of $\mathcal{C}_R$ as possible with the cone of additively reproducible colors. It is clear from the illustration why red, green and blue are used as primaries in televisions and computer monitors. In practice, the weights are between 0 and 1, so the reproducible colors are limited to a finite region as shown in Fig. 3.8(b).

This discussion is easily extended to additive combination with positive coefficients of more than three base colors. In this case, the set of physically reproducible colors lies in a cone with a polygonal cross section where each vertex of the polygon is specified by one of the given colors. However, it may be that some of the base colors lie within the polygon, for example, if white is used as a base color to get a brighter display.

3.11 INDENTIFICATION OF PRIMARIES

Based on the above discussions, we can immediately identify four methods that are frequently used to specify a set of primaries $\mathcal{B} = \{[\mathbf{P}_i], i = 1, 2, 3\}$.

3.11.1 NEW PRIMARIES SPECIFIED IN TERMS OF EXISTING PRIMARIES

The representation of each of the primaries as a linear combination of the XYZ primaries is specified directly, as in equation (3.35), where $[\mathbf{X}]$, $[\mathbf{Y}]$ and $[\mathbf{Z}]$ take the role of the $[\tilde{\mathbf{P}}_i]$. In other words, the

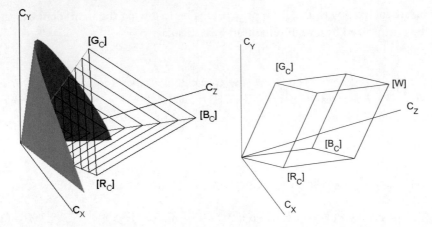

Figure 3.8: (a) Illustration of the cone of reproducible colors with red, green and blue display primaries, in the XYZ color space. (b) The set of reproducible colors with red, green and blue coefficients between 0 and 1.

matrix $\mathbf{A}^{T}_{\mathcal{B}\rightarrow\mathcal{XYZ}}$ is specified. For example, the 1976 CIE Uniform Chromaticity Scale ($\mathcal{UCS}$76) primaries are given by

$$[\mathbf{U'}] = 2.25[\mathbf{X}] + 2.25[\mathbf{Z}]$$
$$[\mathbf{V'}] = [\mathbf{Y}] - 2[\mathbf{Z}] \tag{3.51}$$
$$[\mathbf{W'}] = 3[\mathbf{Z}]$$

and so we have

$$\mathbf{A}_{\mathcal{UCS}76\rightarrow\mathcal{XYZ}} = \begin{bmatrix} 2.25 & 0 & 0 \\ 0 & 1 & 0 \\ 2.25 & -2 & 3 \end{bmatrix}.$$

3.11.2 MATRIX FOR TRANSFORMATION OF TRISTIMULUS VALUES SPECIFIED

The matrix equation to calculate the tristimulus values of an arbitrary color with respect to the primaries $[\mathbf{P}_i]$ as a function of the tristimulus values for the XYZ primaries is specified directly, i.e., the matrix $\mathbf{A}_{\mathcal{XYZ}\rightarrow\mathcal{B}}$ is specified. For the same example of the 1976 UCS primaries with $\mathbf{A}_{\mathcal{UCS}76\rightarrow\mathcal{XYZ}}$ as above, we have

$$\begin{bmatrix} C_{U'} \\ C_{V'} \\ C_{W'} \end{bmatrix} = \mathbf{A}^{-1}_{\mathcal{UCS}76\rightarrow\mathcal{XYZ}} \begin{bmatrix} C_X \\ C_Y \\ C_Z \end{bmatrix}$$
$$= \begin{bmatrix} \frac{4}{9} & 0 & 0 \\ 0 & 1 & 0 \\ -\frac{1}{3} & \frac{2}{3} & \frac{1}{3} \end{bmatrix} \begin{bmatrix} C_X \\ C_Y \\ C_Z \end{bmatrix}. \tag{3.52}$$

3.11.3 SPECTRAL DENSITIES OF PRIMARIES SPECIFIED

The spectral densities $P_i(\lambda)$ of one member of each of the equivalence classes $[\mathbf{P}_i]$ is provided. For example, this could be the spectral density of the light emitted from the red, green and blue phosphors of a CRT display used previously as an example. In this case, the XYZ tristimulus values of each of the primaries can be computed using equation (3.9) with the known XYZ color matching functions of Fig. 3.6, and normalizing them so that their sum matches the reference white $[\mathbf{D}_{65}]$. Applying this, we find

$$
\begin{bmatrix} [\mathbf{R}_C] \\ [\mathbf{G}_C] \\ [\mathbf{B}_C] \end{bmatrix} = \mathbf{A}^T_{\mathcal{RGB}-\mathrm{CRT}\rightarrow\mathcal{XYZ}} \begin{bmatrix} [\mathbf{X}] \\ [\mathbf{Y}] \\ [\mathbf{Z}] \end{bmatrix}
$$
$$
= \begin{bmatrix} 0.4641 & 0.2596 & 0.0358 \\ 0.3056 & 0.6594 & 0.1423 \\ 0.1808 & 0.0810 & 0.9107 \end{bmatrix} \begin{bmatrix} [\mathbf{X}] \\ [\mathbf{Y}] \\ [\mathbf{Z}] \end{bmatrix} .
\tag{3.53}
$$

In fact, the matrix found here was used to determine the CRT RGB color matching functions from the 1931 XYZ color matching functions for the plot of Fig. 3.4.

3.11.4 COLOR MATCHING FUNCTIONS OF NEW PRIMARIES SPECIFIED

Given a color space $\mathcal{C}$ with known visual subspace $\mathcal{VS}_\mathcal{C}$ specified by a particular basis $\mathcal{B}$ and its associated color matching functions, a set of three color matching functions belonging to $\mathcal{VS}_\mathcal{C}$ is provided. This, of course, specifies a new set of primaries. As an example, let $\mathcal{C}_{\mathrm{SB}10}$ be the color space specified by the Stiles and Burch 10° RGB color matching functions [49]. The primaries are $[\mathbf{R}_{\mathrm{SB}10}] = [\delta(\lambda - 645.16)]$, $[\mathbf{G}_{\mathrm{SB}10}] = [\delta(\lambda - 526.32)]$, $[\mathbf{B}_{\mathrm{SB}10}] = [\delta(\lambda - 444.44)]$, and the color matching functions are shown in Fig. 3.9(a) (data from [51]). A set of color matching functions corresponding to the long, medium and short wavelength (LMS) cone photoreceptors of the human retina were given by Stockman and Sharpe [50] in terms of the above RGB basis for $\mathcal{C}_{\mathrm{SB}10}$. These color matching functions, each normalized to have a maximum value of 1, are given by

$$
\begin{bmatrix} \bar{l}_{\mathrm{SS}10}(\lambda) \\ \bar{m}_{\mathrm{SS}10}(\lambda) \\ \bar{s}_{\mathrm{SS}10}(\lambda) \end{bmatrix} = \begin{bmatrix} 0.192413 & 0.749892 & 0.067603 \\ 0.019249 & 0.941916 & 0.113952 \\ 0.0 & 0.010580 & 0.998078 \end{bmatrix} \begin{bmatrix} \bar{r}_{\mathrm{SB}10}(\lambda) \\ \bar{g}_{\mathrm{SB}10}(\lambda) \\ \bar{b}_{\mathrm{SB}10}(\lambda) \end{bmatrix}
\tag{3.54}
$$

and are illustrated in Fig. 3.9(b). Thus, referring to Table 3.1, we identify

$$
\mathbf{A}_{\mathcal{SB}10\rightarrow\mathcal{SS}10} = \begin{bmatrix} 0.192413 & 0.749892 & 0.067603 \\ 0.019249 & 0.941916 & 0.113952 \\ 0.0 & 0.010580 & 0.998078 \end{bmatrix}
\tag{3.55}
$$

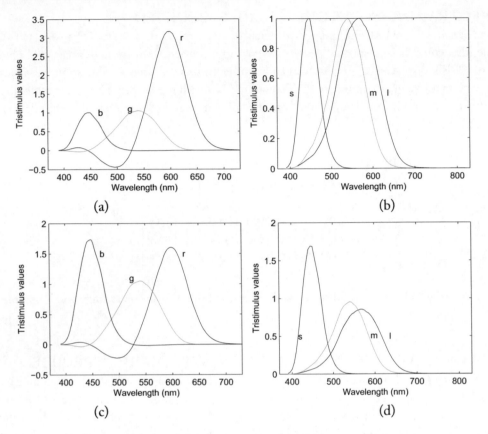

Figure 3.9: Original color matching functions of the (a) Stiles and Burch $10°$ primaries and the (b) Stockman and Sharpe LMS primaries derived from them. (c), (d) Color matching functions of the primaries of (a) and (b), respectively, scaled so that they sum to $[\mathbf{D}_{65}]$.

where the basis of $\mathcal{C}_{SB10}$ corresponding to these color matching functions is denoted $\mathcal{SS}10$. Thus, the LMS basis vectors (primaries) are given by

$$
\begin{bmatrix} [\mathbf{L}_{SS10}] \\ [\mathbf{M}_{SS10}] \\ [\mathbf{S}_{SS10}] \end{bmatrix} = \mathbf{A}_{\mathcal{SB}10\to\mathcal{SS}10}^{-T} \begin{bmatrix} [\mathbf{R}_{SB10}] \\ [\mathbf{G}_{SB10}] \\ [\mathbf{B}_{SB10}] \end{bmatrix}
$$

$$
= \begin{bmatrix} 5.647064 & -0.115554 & 0.001225 \\ -4.497295 & 1.155056 & -0.012244 \\ 0.130966 & -0.124048 & 1.003240 \end{bmatrix} \begin{bmatrix} [\mathbf{R}_{SB10}] \\ [\mathbf{G}_{SB10}] \\ [\mathbf{B}_{SB10}] \end{bmatrix}
\tag{3.56}
$$

Neither of the sets of primaries satisfies the white normalizing convention of Eq. (3.8). Suppose that we choose new bases $\mathcal{SB}10'$ and $\mathcal{SS}10'$ so that a sum of one unit of each primary matches the CIE reference white $[\mathbf{D}_{65}]$ for both bases. It is instructive to carry through this example in detail.

The power density spectrum $D_{65}(\lambda)$ (available at [51]) is shown in Fig. 3.10. Estimating the SB10 tristimulus values using Eq. (3.9), we obtain

$$[\mathbf{D}_{65}] = k(0.555794[\mathbf{R}_{SB10}] + 0.280882[\mathbf{G}_{SB10}] + 0.163324[\mathbf{B}_{SB10}]) \qquad (3.57)$$

where k is an arbitrary constant related to the absolute power of the reference white. For the normalized primaries, the three tristimulus values should be equal, so we arbitrarily choose to scale just $[\mathbf{R}_{SB10}]$ and $[\mathbf{B}_{SB10}]$ for a specific value of k to achieve this, giving

$$\begin{aligned}[\mathbf{D}_{65}] &= 0.280882k(1.978749[\mathbf{R}_{SB10}] + [\mathbf{G}_{SB10}] + 0.581471[\mathbf{B}_{SB10}]) \\ &= [\mathbf{R}'_{SB10}] + [\mathbf{G}'_{SB10}] + [\mathbf{B}'_{SB10}]\end{aligned} \qquad (3.58)$$

with $k = 3.560219$. We can thus express the normalized primaries by

$$\begin{bmatrix} [\mathbf{R}'_{SB10}] \\ [\mathbf{G}'_{SB10}] \\ [\mathbf{B}'_{SB10}] \end{bmatrix} = \begin{bmatrix} 1.978749 & 0 & 0 \\ 0 & 1 & 0 \\ 0 & 0 & 0.581471 \end{bmatrix} \begin{bmatrix} [\mathbf{R}_{SB10}] \\ [\mathbf{G}_{SB10}] \\ [\mathbf{B}_{SB10}] \end{bmatrix}. \qquad (3.59)$$

Referring to Table 3.1, we find

$$\begin{bmatrix} \bar{r}'_{SB10}(\lambda) \\ \bar{g}'_{SB10}(\lambda) \\ \bar{b}'_{SB10}(\lambda) \end{bmatrix} = \begin{bmatrix} 0.505370 & 0 & 0 \\ 0 & 1 & 0 \\ 0 & 0 & 1.719777 \end{bmatrix} \begin{bmatrix} \bar{r}_{SB10}(\lambda) \\ \bar{g}_{SB10}(\lambda) \\ \bar{b}_{SB10}(\lambda) \end{bmatrix} \qquad (3.60)$$

illustrated in Fig. 3.9(c). Combining Eq. (3.56) and Eq. (3.59), we obtain

$$[\mathbf{D}_{65}] = 1.169939[\mathbf{L}_{SS10}] + 1.046265[\mathbf{M}_{SS10}] + 0.590933[\mathbf{S}_{SS10}] \qquad (3.61)$$

thus identifying $[\mathbf{L}'_{SS10}]$, $[\mathbf{M}'_{SS10}]$ and $[\mathbf{S}'_{SS10}]$. The resulting color matching functions are shown in Fig. 3.9(d). Finally, we can write the desired result

$$\begin{aligned}\begin{bmatrix} [\mathbf{L}'_{SS10}] \\ [\mathbf{M}'_{SS10}] \\ [\mathbf{S}'_{SS10}] \end{bmatrix} &= \begin{bmatrix} 1.169939 & 0 & 0 \\ 0 & 1.046265 & 0 \\ 0 & 0 & 0.590933 \end{bmatrix} \begin{bmatrix} [\mathbf{L}_{SS10}] \\ [\mathbf{M}_{SS10}] \\ [\mathbf{S}_{SS10}] \end{bmatrix} \\ &= \begin{bmatrix} 3.338838 & -0.135192 & 0.002464 \\ -2.377950 & 1.208495 & -0.022030 \\ 0.039112 & -0.073304 & 1.019566 \end{bmatrix} \begin{bmatrix} [\mathbf{R}'_{SB10}] \\ [\mathbf{G}'_{SB10}] \\ [\mathbf{B}'_{SB10}] \end{bmatrix}\end{aligned} \qquad (3.62)$$

Note that this result is independent of the actual radiance level of the reference white.

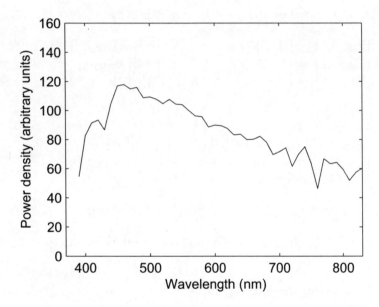

Figure 3.10:   Power density spectrum $D_{65}(\lambda)$ representative of CIE standard illuminant [$\mathbf{D}_{65}$].

CHAPTER 4

Subspaces and Decompositions of the Human Color Space

4.1 INTRODUCTION

So far, we have developed the human color space as a vector space of equivalence classes of spectral densities. Any three linearly independent vectors can form a basis, and several bases have been introduced, including the standardized CIE bases and bases related to certain display devices. In this chapter, a number of further decompositions of the color space will be explored. The first will be based on luminance a measure of relative brightness. This will lead to a decomposition of color space as a direct sum of the two-dimensional subspace of colors of zero luminance and a one-dimensional subspace with luminance as the coordinate. We will then explore a decomposition into equivalence classes that are straight lines through the origin, characterized by chromaticity. We will finish by investigating subspaces induced by certain forms of inherited color deficiency. As in the preceding chapter, a number of axioms are postulated that hold under certain 'normal' viewing conditions for a wide class of human observers. The axioms are then developed in terms of the mathematical theory of color spaces, but of course, conclusions are only valid to the extent that the axioms hold!

4.2 LUMINANCE AND ASSOCIATED DECOMPOSITIONS OF THE COLOR VECTOR SPACE

Luminance is a measure of relative brightness. If two lights have equal luminance, they appear to be equally bright to a viewer, independently of their chromatic attributes. Although it may be difficult to judge if, say, a red light and a blue light have equal brightness when viewing them side by side, this judgement is easier if they are viewed in alternation one after the other. At a low frequency of switching, we can see the display flipping red, blue, red, blue, …. However, as the switching frequency increases and passes a certain limit, the two colors merge into one, which flickers if they have different perceptual brightness. The intensity of one of the lights can be adjusted until the flickering disappears. At this point, the two lights are said to have equal perceptual brightness. This brightness depends on the power density spectrum of the light. A light with a spectrum concentrated near 550 nm appears brighter than a light of equal total power with a spectrum concentrated near 700 nm.

We now state the axioms of brightness matching that will be developed, similar to those of Chapter 3. These were also addressed by Grassmann in his classic paper [16]. The axioms and the conditions under which they hold are discussed at length in [61]. If two lights with spectral densities

$C_1(\lambda), C_2(\lambda) \in \mathcal{P}$ *appear equally bright,* according to a given experimental methodology such as the one discussed above, we say that

$$C_1(\lambda) \mathbin{\triangle} C_2(\lambda). \tag{4.1}$$

We assume that the experimental methodology is such that we can assure that $C(\lambda) \mathbin{\triangle} C(\lambda)$ and that $C_1(\lambda) \mathbin{\triangle} C_2(\lambda)$ implies $C_2(\lambda) \mathbin{\triangle} C_1(\lambda)$. We number the following axioms L1-L4 in analogy to G1-G4 in Chapter 3.

Transitivity (L1). If $C_1(\lambda) \mathbin{\triangle} C_2(\lambda)$ and $C_2(\lambda) \mathbin{\triangle} C_3(\lambda)$, then $C_1(\lambda) \mathbin{\triangle} C_3(\lambda)$.

Coupled with the preceding two conditions, we conclude that $\triangle$ is an equivalence relation on $\mathcal{P}$. We denote the equivalence class of all elements of $\mathcal{P}$ that are a brightness match for $C(\lambda)$ by

$$[C(\lambda)]_\triangle = \{C_1(\lambda) \in \mathcal{P} \mid C_1(\lambda) \mathbin{\triangle} C(\lambda)\}. \tag{4.2}$$

As usual, the set of these equivalence classes forms a partition of $\mathcal{P}$.

The following two empirical facts show that we can scale and add these equivalence classes.

Scaling (L2). If $C_1(\lambda) \mathbin{\triangle} C_2(\lambda)$ then $\alpha C_1(\lambda) \mathbin{\triangle} \alpha C_2(\lambda)$ for any real, nonnegative α.

Thus if $[C(\lambda)]_\triangle$ is as defined in Eq. (4.2), then $[\alpha C(\lambda)]_\triangle = \{\alpha C_1(\lambda) \in \mathcal{P} \mid C_1(\lambda) \mathbin{\triangle} C(\lambda)\}$, which we denote $\alpha[C(\lambda)]_\triangle$ without ambiguity.

Addition (L3). $C(\lambda) + C_1(\lambda) \mathbin{\triangle} C(\lambda) + C_2(\lambda)$ for arbitrary $C(\lambda)$ if and only if $C_1(\lambda) \mathbin{\triangle} C_2(\lambda)$.

Corollary to L3. *If $C_1(\lambda) \mathbin{\triangle} C_2(\lambda)$, then $C_1(\lambda) + C_3(\lambda) \mathbin{\triangle} C_2(\lambda) + C_4(\lambda)$ if and only if $C_3(\lambda) \mathbin{\triangle} C_4(\lambda)$.*

The proof is the same as the proof of the Corollary to G3.

Property L3 allows us to define the addition of brightness classes, i.e., $[C_1(\lambda)]_\triangle + [C_2(\lambda)]_\triangle = [C_1(\lambda) + C_2(\lambda)]_\triangle$, independently of the specific choices of $C_1(\lambda)$ and $C_2(\lambda)$ in the respective classes. The set $\mathcal{L}_P = \{[C(\lambda)]_\triangle \mid C(\lambda) \in \mathcal{P}\}$ with the operation of addition defined above, denoted $(\mathcal{L}_P, +)$, is again a commutative semigroup with a neutral element $[0]_\triangle$ in which every element is cancelable. As in Chapter 3, we extend brightness equivalence to all of $\mathcal{A}$, making the set of equivalence classes a vector space, so as to determine the implications on the color vector space $\mathcal{C}$. Because the algebraic structure of $\mathcal{L}_P$ is the same as that of $\mathcal{C}_P$, the development and proofs are identical to those presented in Chapter 3. Thus, the proofs will be omitted. Only the dimension of the resulting vector space is different.

We define the following relation on $\mathcal{A}$. For $C_1(\lambda), C_2(\lambda) \in \mathcal{A}$,

$$C_1(\lambda) \boxminus C_2(\lambda) \quad \text{if and only if} \quad C_1^+(\lambda) + C_2^-(\lambda) \mathbin{\triangle} C_2^+(\lambda) + C_1^-(\lambda). \tag{4.3}$$

Using the same proofs as in Chapter 3, we determine that $\boxminus$ is an equivalence relation on $\mathcal{A}$ and that properties L2 and L3 apply to $\boxminus$ on $\mathcal{A}$; we denote the extended properties (L2') and (L3'). Thus $\mathcal{L} = \{[C(\lambda)]_\boxminus \mid C(\lambda) \in \mathcal{A}\}$ forms a vector space.

The following axiom is analogous to G4.

Dimension (L4). For any two non-zero lights $C_1(\lambda)$, $C_2(\lambda) \in \mathcal{P}$, we can find $\alpha > 0$ such that $\alpha C_1(\lambda) \triangle C_2(\lambda)$.

This is also easily extended to $\mathcal{A}$, from which we conclude that the vector space $\mathcal{L}$ is of dimension 1. Thus, any non-zero element of $\mathcal{L}$ can be used as a basis vector, and a single scalar coordinate is all that is needed to identify elements of $\mathcal{L}$. Not unexpectedly, this coordinate is computed using a linear functional from $\mathcal{A}^*$, according to the following theorem.

Theorem 4.1   *Let $P(\lambda)$ be any fixed element of $\mathcal{A}$. Then for any arbitrary $C(\lambda) \in \mathcal{A}$, $[C(\lambda)]_\boxdot = C_{LP}[P(\lambda)]_\boxdot$ where $C_{LP} = k \int_\mathcal{V} V(\lambda)C(\lambda)\,d\lambda$. The function $V(\lambda)$ has the properties that $\max_{\lambda \in \mathcal{V}} V(\lambda) = 1$ and*

$$[\delta(\lambda - \mu)]_\boxdot = V(\mu)[\delta(\lambda - \lambda_0)]_\boxdot, \quad where \quad V(\lambda_0) = 1. \tag{4.4}$$

Proof. The proof follows the same argument as Theorem 3.10. Following the informal presentation of that proof, we write $C(\lambda) = C_c(\lambda) + C_d(\lambda)$ and express $C_c(\lambda) = \int_\mathcal{V} C_c(\mu)\delta(\lambda - \mu)\,d\mu$. We define $V_{PL}(\lambda)$ by experimentation over the spectrum of monochromatic lights to satisfy $\delta(\lambda - \mu) \boxdot V_{PL}(\mu)P(\lambda)$. By linearity, $C_c(\lambda) \boxdot \left(\int_\mathcal{V} V_{PL}(\mu)C_c(\mu)\,d\mu\right) P(\lambda)$. Now let $\lambda_0 = \arg\max_\lambda V_{PL}(\lambda)$. Then $[P(\lambda)]_\boxdot = \frac{1}{V_{PL}(\lambda_0)}[\delta(\lambda - \lambda_0)]_\boxdot$. It follows that $[\delta(\lambda - \mu)]_\boxdot = V(\mu)[\delta(\lambda - \lambda_0)]_\boxdot$ where $V(\lambda) = \frac{V_{PL}(\lambda)}{V_{PL}(\lambda_0)}$, which of course, has a maximum value of 1. The result applies directly to $C_d(\lambda)$ and thus to $C(\lambda)$. Because of the continuity of $V(\lambda)$, the limits behind this informal proof are well behaved □

The function $V(\lambda)$ is called the *spectral luminous efficiency* and was standardized by the CIE in 1924. We assume normal daylight vision involving the cones, in which case $V(\lambda)$ refers to the standard *photopic* observer. Standardized values are given from 360nm to 830nm at one-nm intervals and are available in the previously mentioned sources [61], [51]. Fig. 4.1 shows the standard photoptic spectral luminous efficiency curve. The radiometric units of Section 2.1 are converted to photometric units using expressions of the form

$$L_v = K_m \int_\mathcal{V} L_e(\lambda)V(\lambda)\,d\lambda \tag{4.5}$$

were $K_m = 683.002$ lm/W (lm stands for lumen). A great more detail is provided in [61]. For simplicity, we refer to any photometric quantity as luminance or relative luminance, even without the factor K_m. Specifically, if a reference white is specified to have a luminance of 1 in normalized units, we refer to the photometric quantity on this scale as *relative luminance*.

The main purpose of this section is to consider the situation when the color matching axioms G1-G4, and the brightness matching axioms L1-L4 apply simultaneously to the same observer. In

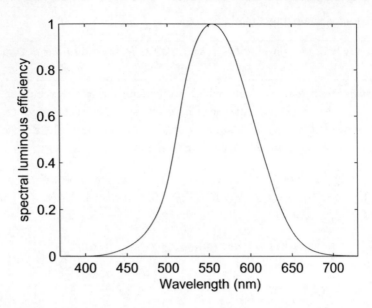

Figure 4.1: Photoptic spectral luminous efficiency curve standardized by the CIE in 1924.

this case, if two lights match colorimetrically, they are identical in all respects including brightness, and so they must also match in brightness:

$$C_1(\lambda) \boxminus C_2(\lambda) \Rightarrow C_1(\lambda) \boxminus C_2(\lambda). \tag{4.6}$$

In this case, we can conclude that $V(\lambda) \in \mathcal{VS}_C$, where C is the color space for the given observer.

Theorem 4.2   *For an observer described by a color space C and a spectral luminous efficiency $V(\lambda)$, $V(\lambda) \in \mathcal{VS}_C$.*

Proof. Although this can be considered to be a special case of Theorem 3.12, it is instructive to provide a detailed proof. Let $\mathcal{B} = \{[\mathbf{P}_1], [\mathbf{P}_2], [\mathbf{P}_3]\}$ be any set of primaries for C with $P_1(\lambda)$, $P_2(\lambda)$, $P_3(\lambda)$ arbitrary elements of the corresponding equivalence classes. Then

$$C(\lambda) \boxminus C_1 P_1(\lambda) + C_2 P_2(\lambda) + C_3 P_3(\lambda) \tag{4.7}$$

and so by Eq. (4.6)

$$C(\lambda) \boxminus C_1 P_1(\lambda) + C_2 P_2(\lambda) + C_3 P_3(\lambda). \tag{4.8}$$

Thus

$$\int_{\mathcal{V}} C(\lambda) V(\lambda)\, d\lambda = C_1 \int_{\mathcal{V}} P_1(\lambda) V(\lambda)\, d\lambda + C_2 \int_{\mathcal{V}} P_2(\lambda) V(\lambda)\, d\lambda + C_3 \int_{\mathcal{V}} P_3(\lambda) V(\lambda)\, d\lambda. \tag{4.9}$$

The terms $P_{iL} = \int_{\mathcal{V}} P_i(\lambda)V(\lambda)\,d\lambda$ describe the luminance of the primaries and are, of course, independent of the specific choices of the $P_i(\lambda)$ from the respective equivalence classes. Thus

$$C_L = C_1 P_{1L} + C_2 P_{2L} + C_3 P_{3L}. \tag{4.10}$$

Now writing $C_i = \int_{\mathcal{V}} \bar{p}_i(\lambda)C(\lambda)\,d\lambda$, we see that *for any* $C(\lambda)$

$$\int_{\mathcal{V}} C(\lambda)V(\lambda)\,d\lambda = \int_{\mathcal{V}} (P_{1L}\bar{p}_1(\lambda) + P_{2L}\bar{p}_2(\lambda) + P_{3L}\bar{p}_3(\lambda))C(\lambda)\,d\lambda \tag{4.11}$$

from which we conclude that

$$V(\lambda) = P_{1L}\bar{p}_1(\lambda) + P_{2L}\bar{p}_2(\lambda) + P_{3L}\bar{p}_3(\lambda) \in \mathcal{VS}_C \tag{4.12}$$

$$\square$$

In the course of the proof, we have also identified the linear mapping from $\mathcal{C} \to \mathcal{L} : [\mathbf{C}] \mapsto C_L$ given in Eq. (4.10).

The CIE XYZ basis that we have already seen was selected in such a way that $\bar{y}(\lambda) = V(\lambda)$, so the the Y tristimulus value also provides the luminance.

The set of all elements of $\mathcal{C}$ with the same luminance forms an equivalence class which can be used for a canonical decomposition of $\mathcal{C}$. Let $\mathcal{L}_0 = \{[\mathbf{C}] \mid C_L = 0\}$. This is a two-dimensional subspace (i.e., a plane) of $\mathcal{C}$ consisting of points satisfying Eq. (4.10) with $C_L = 0$. Thus there exists a (non-unique) subspace $\mathcal{L}_1$ of dimension 1 such that

$$\mathcal{C} = \mathcal{L}_0 \oplus \mathcal{L}_1. \tag{4.13}$$

Under this decomposition, any element of $\mathcal{C}$ can be written uniquely as $[\mathbf{C}] = [\mathbf{C}_0] + [\mathbf{C}_1]$ where $[\mathbf{C}_0] \in \mathcal{L}_0$ and $[\mathbf{C}_1] \in \mathcal{L}_1$. The coefficient of the component in $\mathcal{L}_1$ is the luminance if we choose the basis vector in $\mathcal{L}_1$ to have luminance 1.

We have already seen two bases for $\mathcal{C}$ that are in this form, namely $\mathcal{XYZ}$ and $\mathcal{UCS}76$. For $\mathcal{XYZ}$, we see that $\mathcal{L}_0 = \mathrm{span}([\mathbf{X}], [\mathbf{Z}])$ and $\mathcal{L}_1 = \mathrm{span}([\mathbf{Y}])$ while for $\mathcal{UCS}76$, $\mathcal{L}_0 = \mathrm{span}([\mathbf{U}'], [\mathbf{W}'])$ and $\mathcal{L}_1 = \mathrm{span}([\mathbf{V}'])$. Since $[\mathbf{U}'], [\mathbf{W}']$ are a linear combination of $[\mathbf{X}]$ and $[\mathbf{Z}]$, $\mathcal{L}_0$ is indeed the same for both bases whereas $\mathcal{L}_1$ is different for the two bases. We could also define a decomposition where $\mathcal{L}_1 = \mathrm{span}([\mathbf{W}])$ where $[\mathbf{W}]$ is a suitable reference white. An example of such a basis, arbitrarily called $\mathcal{W}0 = \{[\mathbf{W}], [\mathbf{Q}_1], [\mathbf{Q}_2]\}$, is defined by the transformation matrix

$$\mathbf{A}_{\mathcal{W}0 \to \mathcal{XYZ}} = \begin{bmatrix} 0.9505 & -0.1661 & 0.1397 \\ 1.0 & 0.0 & 0.0 \\ 1.0891 & 0.4013 & 0.4809 \end{bmatrix} \tag{4.14}$$

Equivalently, using Table 3.1, the new basis vectors (primaries) are

$$\begin{aligned}
[\mathbf{W}] &= 0.9505[\mathbf{X}] + [\mathbf{Y}] + 1.0891[\mathbf{Z}] = [\mathbf{D}_{65}] \\
[\mathbf{Q}_1] &= -0.1661[\mathbf{X}] + 0.4013[\mathbf{Z}] \\
[\mathbf{Q}_2] &= 0.1397[\mathbf{X}] + 0.4809[\mathbf{Z}]
\end{aligned} \tag{4.15}$$

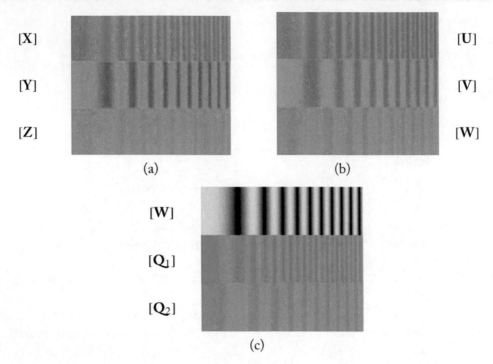

Figure 4.2: Illustration of three bases for color space where one tristimulus value is relative luminance. (a) XYZ. (b) CIE 1976 UCS. (c)Basis $\mathcal{W}0$ defined in the text.

In this basis (which I have devised myself), one basis vector is reference white $[\mathbf{D}_{65}]$ while the other two are in the plane of zero luminance spanned by $[\mathbf{X}]$ and $[\mathbf{Z}]$. They were chosen so that $[\mathbf{Q}_2]$ appears to have the smallest visual bandwidth and $[\mathbf{Q}_1]$ the largest. It is given here simply as an example, and it is similar in spirit to the YIQ representation used in NTSC television.

It is interesting to visualize these three bases. Fig. 4.2 shows a test pattern for each consisting of three horizontal stripes. Only one tristimulus value varies in each strip along the horizontal axis about a gray value, $[\mathbf{C}_i](x) = [\mathbf{GR}] + f(x)[\mathbf{P}_i], i = 1, 2, 3$. Here, $f(x)$ is a sinusoidal type signal; the exact form is not important. Note that all three components carry chromatic information in the XYZ and UCS bases whereas only two carry chromatic information in the proposed basis. One tristimulus value is relative luminance in each of the bases.

4.3   CHROMATICITY CLASSES

A very popular decomposition for color spaces involves the use of chromaticity classes. A chromaticity class simply consists of all scalar multiples of a given color $[\mathbf{C}]$, namely $\{\alpha[\mathbf{C}] \mid \alpha \in \mathbb{R}_+^*\}$. If $[\mathbf{C}]$ is physically realizable, this is a straight line within the cone $\mathcal{C}_R$. For a given light with spectral density $C(\lambda)$, the chromaticity class includes all multiples $\alpha C(\lambda)$, for $\alpha > 0$. These curves share a common 'chromatic' attribute, but vary in their brightness. A chromaticity class is easily seen to be an equivalence class. The standard procedure to identify a particular chromaticity class with respect to a given basis $\mathcal{B} = \{[\mathbf{P}_1], [\mathbf{P}_2], [\mathbf{P}_3]\}$ is to give its intersection with the plane $C_1 + C_2 + C_3 = 1$. Thus, if $[\mathbf{C}] = C_1[\mathbf{P}_1] + C_2[\mathbf{P}_2] + C_3[\mathbf{P}_3]$ is an arbitrary color, we seek α such that the tristimulus values of $\alpha[\mathbf{C}]$ lie on this plane. This is achieved if $\alpha = 1/(C_1 + C_2 + C_3)$, with corresponding tristimulus values

$$c_i = \frac{C_i}{C_1 + C_2 + C_3}. \tag{4.16}$$

This will have a solution if $[\mathbf{C}]$ lies in the half space given by $C_1 + C_2 + C_3 > 0$. These are called *chromaticity coordinates*, usually denoted with the same symbol as tristimulus values, but in lowercase.

Since only two chromaticity values are needed ($c_3 = 1 - c_1 - c_2$), chromaticities are generally plotted on a two-dimensional diagram with axes c_1 and c_2. The chromaticity diagram for the $\mathcal{XYZ}$ basis is shown in Fig. 4.3. The spectral locus lies along the shark-fin shaped boundary; a number of specific wavelengths within $\mathcal{V}$ are indicated. The line joining the extreme spectral colors is called the line of purples. This diagram shows the intersection of the cone $\mathcal{C}_R$ (see Fig. 3.7) with the plane $C_X + C_Y + C_Z = 1$, projected on the plane $C_Z = 0$. All physically realizable colors have chromaticities that lie with the region bounded by the spectral locus and the line of purples. We note that if the chosen primaries were normalized with respect to a reference white, $[\mathbf{W}] = [\mathbf{P}_1] + [\mathbf{P}_2] + [\mathbf{P}_3]$, the chromaticity coordinates of the reference white are $(\frac{1}{3}, \frac{1}{3})$.

The set of chromaticity classes forms a projective space. If we add two colors, the tristimulus values are simply added. A more complex computation is required to compute the new chromaticities.

4.3.1 DETERMINATION OF TRISTIMULUS VALUES FROM LUMINANCE AND CHROMATICITIES

Suppose that for a given color $[\mathbf{C}]$, we are given the luminance C_L and the chromaticities c_1 and c_2 with respect to some set of primaries $[\mathbf{P}_1]$, $[\mathbf{P}_2]$ and $[\mathbf{P}_3]$ (of course $c_3 = 1 - c_1 - c_2$). We know the primaries, so the luminosity coefficients P_{1L}, P_{2L}, P_{3L} are known. We want to find the tristimulus values C_1, C_2, C_3.

Solution: We know that $C_L = C_1 P_{1L} + C_2 P_{2L} + C_3 P_{3L}$ (Eq. (4.10)). Dividing by $C_1 + C_2 + C_3$, we obtain

$$\frac{C_L}{C_1 + C_2 + C_3} = c_1 P_{1L} + c_2 P_{2L} + c_3 P_{3L}.$$

Now multiplying by C_i for $i = 1, 2, 3$ yields

$$C_L c_i = C_i (c_1 P_{1L} + c_2 P_{2L} + c_3 P_{3L})$$

and thus

$$C_i = \frac{C_L c_i}{c_1 P_{1L} + c_2 P_{2L} + c_3 P_{3L}}, \quad i = 1, 2, 3. \tag{4.17}$$

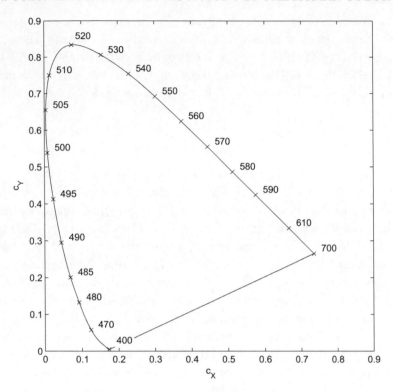

Figure 4.3: Chromaticity diagram for CIE 1931 XYZ primaries.

These expressions are particularly simple for the XYZ basis, since $X_L = Z_L = 0$ and $Y_L = 1$. It follows that

$$C_X = \frac{C_L c_X}{c_Y}$$
$$C_Y = C_L \tag{4.18}$$
$$C_Z = \frac{C_L c_Z}{c_Y}.$$

4.3.2 ADDITIVE REPRODUCTION OF COLORS REVISITED

We have seen in Section 3.10.2 that any color [**Q**] that is formed as the linear combination of three physically realizable colors, [**A**], [**B**] and [**C**], with non-negative coefficients, lies within a convex cone with a triangular cross section (Fig. 3.8). The three colors lie on the edges of this cone. It is clear that the intersection of this cone with the plane $Q_1 + Q_2 + Q_3 = 1$ is a triangle, and its projection on the plane $Q_3 = 0$ is also a triangle, with the three colors at the vertices. Although this argument may be sufficient, the following gives an explicit proof, showing the manipulation of chromaticities in the projective space.

Theorem 4.3 *On a chromaticity diagram, the chromaticities of all colors that are realized by the sum of a positive quantity of three physical colors lie within a triangle whose vertices are the chromaticities of these three colors.*

Proof. We first show that the chromaticities of all colors that are realized by the sum of a positive quantity of *two* physical colors lie on the straight line joining the chromaticities of the two colors. Then, adding this sum to a positive quantity of the third color will yield chromaticities that lie within the stated triangle. To understand why the chromaticities of a sum of two colors lie on the straight line joining their chromaticities, we observe that in the three-dimensional color space, the sum of two colors lies on the plane that contains the lines through the origin and each of these two colors. The chromaticities of the sum lie on the intersection of this plane and the plane $Q_1 + Q_2 + Q_3 = 1$, which is a straight line. The projection of this straight line on the chromaticity diagram is also a straight line.

More formally, let the primaries with respect to which we are computing chromaticities be $[\mathbf{P}_1]$, $[\mathbf{P}_2]$ and $[\mathbf{P}_3]$, and the three physical colors under consideration be $[\mathbf{A}]$, $[\mathbf{B}]$ and $[\mathbf{C}]$. Thus, $[\mathbf{A}] = A_1[\mathbf{P}_1] + A_2[\mathbf{P}_2] + A_3[\mathbf{P}_3]$ and $a_i = A_i/(A_1 + A_2 + A_3)$, with similar expressions for $[\mathbf{B}]$ and $[\mathbf{C}]$. Now let $[\mathbf{Q}] = \alpha_1[\mathbf{A}] + \alpha_2[\mathbf{B}]$ where $\alpha_1, \alpha_2 > 0$, so that $Q_i = \alpha_1 A_i + \alpha_2 B_i$ for $i = 1, 2, 3$. Then the chromaticities of this mixture are given by

$$q_i = \frac{\alpha_1 A_i + \alpha_2 B_i}{\alpha_1(A_1 + A_2 + A_3) + \alpha_2(B_1 + B_2 + B_3)}. \tag{4.19}$$

Cross-multiplying this equation, and substituting $A_i = a_i(A_1 + A_2 + A_3)$ and $B_i = b_i(B_1 + B_2 + B_3)$, gives

$$\alpha_1(A_1 + A_2 + A_3)a_i + \alpha_2(B_1 + B_2 + B_3)b_i = (\alpha_1(A_1 + A_2 + A_3) + \alpha_2(B_1 + B_2 + B_3))q_i \tag{4.20}$$

or rearranging

$$\alpha_1(A_1 + A_2 + A_3)(a_i - q_i) + \alpha_2(B_1 + B_2 + B_3)(b_i - q_i) = 0, \quad i = 1, 2, 3. \tag{4.21}$$

Now, using similar triangles, we observe that the segment of the straight line joining (a_1, a_2) and (b_1, b_2) is described by

$$\frac{q_1 - a_1}{b_1 - q_1} = \frac{q_2 - a_2}{b_2 - q_2} = \gamma > 0 \tag{4.22}$$

or equivalently

$$(a_i - q_i) + \gamma(b_i - q_i) = 0, \quad i = 1, 2. \tag{4.23}$$

The conclusion follows. □

This is illustrated in the xy chromaticity diagram of Fig. 4.4. The chromaticities of red, green and blue primaries typical of the phosphors of a modern CRT are shown. The subset of all possible colors that can be reproduced on this CRT have chromaticities that lie within the indicated triangle. Each point is illustrated by the color on the boundary of the set of reproducible colors (Fig. 3.8(b)) with the given chromaticity. This explains why red, green and blue are used as primaries in additive color reproduction systems.

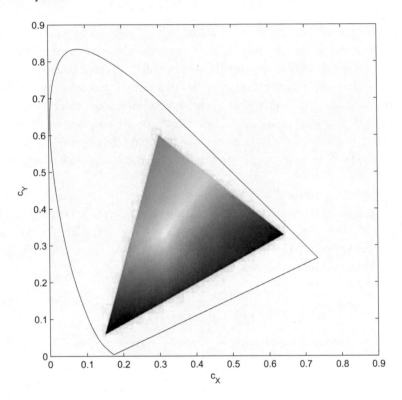

Figure 4.4:   Chromaticities of colors that can be reproduced by addition of positive quantities of three primaries.

4.4   DECOMPOSITION OF COLOR SPACE CORRESPONDING TO CERTAIN COLOR DEFICIENCIES

There are many types of deficiencies in color vision in the human population, including certain inherited types. In this section, we consider those that result in a two-dimensional color space in which all color matches made by a 'normal' observer with a three-dimensional color space are accepted. Observers for which the dimension of the color space is two are called dichromats (as opposed to trichromats when the dimension is three). There are two main types, protanopes and deuteranopes, and a third rare type, tritanopes. Each of these types of deficiency leads to a certain canonical decomposition of color space, and these decompositions can be combined to yield a single decomposition that reflects the three types of dichromatic color vision. This decomposition leads to one procedure to estimate the spectral sensitivities of the L, M and S cones of a normal trichromat, as will be shown later in the section.

Let $\mathcal{D}$ be the color space of any one of the types of dichromats, assumed to be of dimension two. Once again, there is a color-matching equivalence on $\mathcal{P}$ which can be extended to an equivalence on $\mathcal{A}$. Everything is identical to the development in Chapter 3 except that the dimension is two rather than three. In other words, the modified version of G4 is that three colors are always linearly dependent, but there exist sets of two linearly independent colors in the dichromat color space. Let the extended dichromatic equivalence be denoted $\boxminus$. The assumption that color matches of a normal viewer are accepted by dichromats means

$$C_1(\lambda) \boxminus C_2(\lambda) \Rightarrow C_1(\lambda) \boxminus C_2(\lambda). \tag{4.24}$$

This implies that the visual subspace of a dichromat $\mathcal{VS}_{\mathcal{D}}$ is a subspace of $\mathcal{VS}_{\mathcal{C}}$. This can be proved in a manner analogous to the proof of Theorem 4.1. However, no proof is needed here, as this is a special case of Theorem 3.12.

Theorem 4.4   *For a dichromat with color space $\mathcal{D}$ that accepts color matches of a trichromat with color space $\mathcal{C}$, $\mathcal{VS}_{\mathcal{D}} \subset \mathcal{VS}_{\mathcal{C}}$.*

By hypothesis, $\dim(\mathcal{VS}_{\mathcal{D}}) = 2$. Suppose that $\bar{q}_1(\lambda)$ and $\bar{q}_2(\lambda)$ span $\mathcal{VS}_{\mathcal{D}}$, corresponding to a basis $\mathcal{E} = \{[\mathbf{Q}_1], [\mathbf{Q}_2]\}$ of $\mathcal{D}$. Also, let $\bar{p}_1(\lambda), \bar{p}_2(\lambda)$, and $\bar{p}_3(\lambda)$ be the color matching functions for $\mathcal{C}$ with respect to a basis $\mathcal{B} = \{[\mathbf{P}_1], [\mathbf{P}_2], [\mathbf{P}_3]\}$. From Theorem 4.4, we can write

$$\begin{bmatrix} \bar{q}_1(\lambda) \\ \bar{q}_2(\lambda) \end{bmatrix} = \begin{bmatrix} a_{11} & a_{12} & a_{13} \\ a_{21} & a_{22} & a_{23} \end{bmatrix} \begin{bmatrix} \bar{p}_1(\lambda) \\ \bar{p}_2(\lambda) \\ \bar{p}_3(\lambda) \end{bmatrix}$$
$$= \mathbf{A}_{\mathcal{B} \to \mathcal{E}} \begin{bmatrix} \bar{p}_1(\lambda) \\ \bar{p}_2(\lambda) \\ \bar{p}_3(\lambda) \end{bmatrix}. \tag{4.25}$$

The matrix $\mathbf{A}_{\mathcal{B} \to \mathcal{E}}$ can be found by experiments with dichromats, of which many have been reported. We can also immediately write

$$\begin{bmatrix} C_{Q1} \\ C_{Q2} \end{bmatrix} = \mathbf{A}_{\mathcal{B} \to \mathcal{E}} \begin{bmatrix} C_{P1} \\ C_{P2} \\ C_{P3} \end{bmatrix}. \tag{4.26}$$

Here we can interpret $\mathbf{A}_{\mathcal{B} \to \mathcal{E}}$ as a matrix representing a linear transformation from $\mathcal{C}$ to $\mathcal{D}$ with respect to the given bases. Denote this transformation $\mathcal{S}_{\mathcal{C} \to \mathcal{D}}$, and let $\mathcal{D}_0 = \ker \mathcal{S}_{\mathcal{C} \to \mathcal{D}}$. By definition, for any $[\mathbf{C}] \in \mathcal{C}$, every point of the coset $[\mathbf{C}] + \mathcal{D}_0$ maps to the same point in $\mathcal{D}$, i.e., all points in this coset, which appear different to a trichromat, appear the same to a dichromat. Thus, we can write $C = \mathcal{D}_0 \oplus \mathcal{D}_1$, where $\mathcal{D}_1$ is isomorphic to $\mathcal{D}$. Note that $\mathcal{D}_0$ is a unique one-dimensional subspace of C, but $\mathcal{D}_1$ is not unique.

Let us call the color spaces for protanopes, deuteranopes, and tritanopes $\mathcal{D}_P$, $\mathcal{D}_D$ and $\mathcal{D}_T$, respectively. Similarly, let $\mathcal{D}_{P0}$, $\mathcal{D}_{D0}$ and $\mathcal{D}_{T0}$ denote the kernel of the respective transformations $\mathcal{S}_{C \to \mathcal{D}_P}$, $\mathcal{S}_{C \to \mathcal{D}_D}$ and $\mathcal{S}_{C \to \mathcal{D}_T}$. Each of these is a one-dimensional subspace of C, and they are independent, so one vector from each forms a basis for C. This independence is an empirical observation on the actual population of dichromats. The following fundamental decomposition of C follows.

Theorem 4.5   *Let* $[\mathbf{Q}_P]$, $[\mathbf{Q}_D]$ *and* $[\mathbf{Q}_T]$ *be nonzero elements of* $\mathcal{D}_{P0}$, $\mathcal{D}_{D0}$ *and* $\mathcal{D}_{T0}$, *respectively. Then* $[\mathbf{Q}_P]$, $[\mathbf{Q}_D]$ *and* $[\mathbf{Q}_T]$ *form a basis for* C *corresponding to the decomposition* $C = \mathcal{D}_{P0} \oplus \mathcal{D}_{D0} \oplus \mathcal{D}_{T0}$. *Let the corresponding reciprocal basis of* VS_C *be* $\bar{q}_P(\lambda)$, $\bar{q}_D(\lambda)$ *and* $\bar{q}_T(\lambda)$. *Then* $\bar{q}_P(\lambda) \in VS_{\mathcal{D}_D} \cap VS_{\mathcal{D}_T}$, $\bar{q}_D(\lambda) \in VS_{\mathcal{D}_P} \cap VS_{\mathcal{D}_T}$ *and* $\bar{q}_T(\lambda) \in VS_{\mathcal{D}_P} \cap VS_{\mathcal{D}_D}$ *leading to the corresponding decomposition of* VS_C

$$VS_C = (VS_{\mathcal{D}_D} \cap VS_{\mathcal{D}_T}) \oplus (VS_{\mathcal{D}_P} \cap VS_{\mathcal{D}_T}) \oplus (VS_{\mathcal{D}_P} \cap VS_{\mathcal{D}_D}).$$

Proof. By hypothesis, the three spaces are different and independent, so that $C = \mathcal{D}_{P0} \oplus \mathcal{D}_{D0} \oplus \mathcal{D}_{T0}$. Let $\mathcal{PDT}$ denote the basis $\{[\mathbf{Q}_P], [\mathbf{Q}_D], [\mathbf{Q}_T]\}$, and let $\mathcal{E}_P$, $\mathcal{E}_D$ and $\mathcal{E}_T$ denote arbitrary bases for $\mathcal{D}_P$, $\mathcal{D}_D$ and $\mathcal{D}_T$, respectively. Then, since $[\mathbf{Q}_P] \in \mathcal{D}_{P0}$, by definition

$$\mathbf{A}_{\mathcal{PDT} \to \mathcal{E}_P} \begin{bmatrix} 1 \\ 0 \\ 0 \end{bmatrix} = \begin{bmatrix} 0 \\ 0 \end{bmatrix} \tag{4.27}$$

i.e., the first column of $\mathbf{A}_{\mathcal{PDT} \to \mathcal{E}_P}$ is zero. Then, from Eq. 4.25, it follows that $VS_{\mathcal{D}_P} = \text{span}(\bar{p}_D(\lambda), \bar{p}_T(\lambda))$. Similarly, $VS_{\mathcal{D}_D} = \text{span}(\bar{p}_P(\lambda), \bar{p}_T(\lambda))$ and $VS_{\mathcal{D}_T} = \text{span}(\bar{p}_P(\lambda), \bar{p}_D(\lambda))$. Since $\bar{p}_P(\lambda)$, $\bar{p}_D(\lambda)$ and $\bar{p}_T(\lambda)$ are linearly independent, the conclusion follows. $\square$

To investigate these ideas numerically, we use the specification of the visual subspaces for the three types of dichromats in terms of the CIE 1931 XYZ basis from [61]. The three transformation matrices are directly obtained from Table 2(5.14.2) of [61]:

$$\mathbf{A}_{\mathcal{XYZ} \to \mathcal{E}_P} = \begin{bmatrix} 0.010597 & -0.031290 & 0.59840 \\ -0.41968 & 1.23912 & 0.028599 \end{bmatrix}$$

$$\mathbf{A}_{\mathcal{XYZ} \to \mathcal{E}_D} = \begin{bmatrix} 0.0 & 0.0 & 0.59909 \\ 0.57870 & 7.8125 & -0.38164 \end{bmatrix} \tag{4.28}$$

$$\mathbf{A}_{\mathcal{XYZ} \to \mathcal{E}_T} = \begin{bmatrix} -0.47012 & 1.2456 & 0.096974 \\ 3.7874 & -0.68891 & -0.78123 \end{bmatrix}$$

The intersection of the visual subspaces can be obtained by finding the intersection of the row spaces of the respective pairs of matrices above. This can be accomplished using the singular value decomposition and algorithm 12.4-3 of [15]. We thus find that

$$
\begin{aligned}
\mathcal{VS}_{\mathcal{D}_P} \cap \mathcal{VS}_{\mathcal{D}_D} &= \mathrm{span}(\bar{z}(\lambda)) \\
\mathcal{VS}_{\mathcal{D}_P} \cap \mathcal{VS}_{\mathcal{D}_T} &= \mathrm{span}(-0.3201\bar{x}(\lambda) + 0.9451\bar{y}(\lambda) + 0.066\bar{z}(\lambda)) \\
\mathcal{VS}_{\mathcal{D}_D} \cap \mathcal{VS}_{\mathcal{D}_T} &= \mathrm{span}(0.0739\bar{x}(\lambda) + 0.9972\bar{y}(\lambda) - 0.0152\bar{z}(\lambda))
\end{aligned}
\tag{4.29}
$$

The resulting three functions, normalized to have a maximum of 1.0, are plotted in Fig. 4.5(d). Note the resemblance to Fig. 3.9(b).

An equivalent approach according to Theorem 4.5 is to directly determine $[\mathbf{Q}_P]$, $[\mathbf{Q}_D]$ and $[\mathbf{Q}_T]$ in the XYZ space by finding the null space of the matrices in Eq. (4.28), and scaling them so that $[\mathbf{Q}_P] + [\mathbf{Q}_D] + [\mathbf{Q}_T] = [\mathbf{D}_{65}]$. Doing this, we find

$$
\begin{aligned}
\begin{bmatrix} [\mathbf{Q}_P] \\ [\mathbf{Q}_D] \\ [\mathbf{Q}_T] \end{bmatrix} &= \mathbf{A}^T_{\mathcal{PDT} \to \mathcal{XYZ}} \begin{bmatrix} [\mathbf{X}] \\ [\mathbf{Y}] \\ [\mathbf{Z}] \end{bmatrix} \\
&= \begin{bmatrix} 2.552957 & 0.864666 & 0.000003 \\ -1.827044 & 0.135336 & 0.000000 \\ 0.224587 & -0.000002 & 1.088797 \end{bmatrix} \begin{bmatrix} [\mathbf{X}] \\ [\mathbf{Y}] \\ [\mathbf{Z}] \end{bmatrix}
\end{aligned}
\tag{4.30}
$$

with the corresponding transformation matrix

$$
\mathbf{A}_{\mathcal{PDT} \to \mathcal{XYZ}} = \begin{bmatrix} 2.552957 & -1.827044 & 0.224587 \\ 0.864666 & 0.135336 & -0.000002 \\ 0.000003 & 0.000000 & 1.088797 \end{bmatrix}.
\tag{4.31}
$$

From this we find

$$
\begin{aligned}
\begin{bmatrix} \bar{p}_P(\lambda) \\ \bar{p}_D(\lambda) \\ \bar{p}_T(\lambda) \end{bmatrix} &= \mathbf{A}^{-1}_{\mathcal{PDT} \to \mathcal{XYZ}} \begin{bmatrix} \bar{x}(\lambda) \\ \bar{y}(\lambda) \\ \bar{z}(\lambda) \end{bmatrix} \\
&= \begin{bmatrix} 0.070294 & 0.948971 & -0.014498 \\ -0.449110 & 1.326012 & 0.092640 \\ -0.000000 & -0.000002 & 0.918445 \end{bmatrix} \begin{bmatrix} \bar{x}(\lambda) \\ \bar{y}(\lambda) \\ \bar{z}(\lambda) \end{bmatrix},
\end{aligned}
\tag{4.32}
$$

which are plotted in Fig. 4.5(e). These are of course just scaled versions of the ones found previously and plotted in Fig. 4.5(d). We thus see how this decomposition of the color space $\mathcal{C}$ leads to a method to estimate the spectral sensitivities of the three types of cones involved in normal trichromatic color vision.

In the spirit of Fig. 4.2, we can illustrate the LMS primaries derived above. Fig. 4.6 shows a test pattern of the same form as Fig. 4.2 for the primaries defined in Eq. (4.31). Note that when

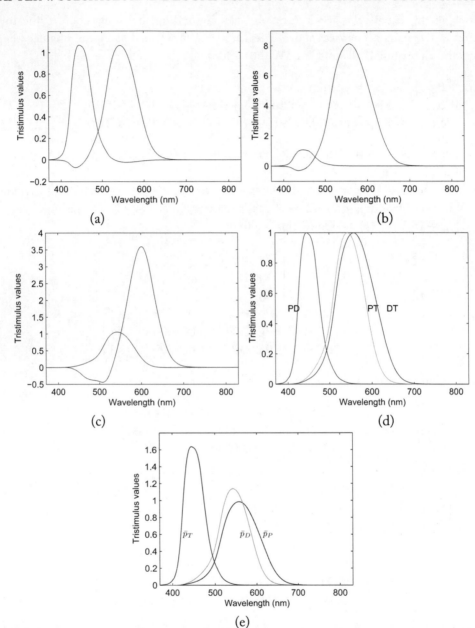

Figure 4.5: Color matching functions of three types of dichromats as obtained from [61]. (a) Protanopes. (b) Deuteranopes. (c) Tritanopes. (d) Color matching functions obtained as the intersections of pairs of visual subspaces as described in the text. (e) Color matching functions determined from basis formed by null spaces as described in the text.

[L]

[M]

[S]

Figure 4.6: Illustration of LMS basis specified in Eq. 4.31.

correctly displayed on a calibrated sRGB monitor, one of the three patterns would appear perfectly flat for a standard dichromat.

We have seen that for any of the standard types of dichromat, the trichromatic color space can be decomposed as a direct sum $\mathcal{C} = \mathcal{D}_0 \oplus \mathcal{D}_1$, where $\mathcal{D}_0$ is a unique one-dimensional subspace of $\mathcal{C}$ that characterizes the type of dichromat, while $\mathcal{D}_1$ is an arbitrary two-dimensional subspace (as long as it does not contain $\mathcal{D}_0$). For any color in $\mathcal{C}$, we can add to it an arbitrary element of $\mathcal{D}_0$ and the appearance of that color will not change for a dichromat. Thus, we could choose an arbitrary two-dimensional manifold in $\mathcal{C}$ subject to the constraint that each line parallel to $\mathcal{D}_0$ intersects it at one point only, and project each point of $\mathcal{C}$ to that manifold along $\mathcal{D}_0$. If this was done for each pixel in an image, the resulting image would appear unchanged to a dichromat of the given type. Brettel, Viénot and colleagues have studied this, to illustrate approximately for a trichromat what a dichromat would perceive [55], [2]. Based on various studies, they have concluded that a manifold consisting of two half planes intersecting along the line through the origin representing equal energy stimuli (neutral) is suitable. The two half planes are fixed by choosing one color on each that is perceived to be of the same hue by normal trichromats and dichromats of the given type. Note that images that appear to be neutral (gray scale) to a trichromat also appear to be neutral to a dichromat. These constraints were based on observers with trichromatic vision in one eye and dichromatic vision in the other. Although there are many caveats on this result, the images are nonetheless instructive. Software has been made publicly available to simulate this projection [5]. I used the Vischeck Photoshop plugin to generate two-dimensional projections of the image of colors on the plane $C_R + C_G + C_B = 1$ for the sRGB primaries as projected in a xy chromaticity diagram (similar to Fig. 4.4). The resulting images are shown in Fig. 4.7.

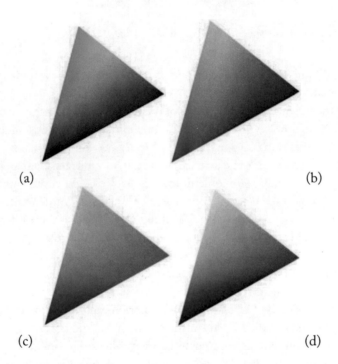

Figure 4.7: Images of the sRGB color triangle in xy chromaticity rendered with the Vischeck Photoshop plugin. (a) Original trichromatic image. (b) Image as perceived by a tritanope. (c) Image as perceived by a deuteranope. (d) Image as perceived by a protanope.

CHAPTER 5

Various Color Spaces, Representations, and Transformations

This chapter introduces several important color spaces, some non-linear transformations of color space, and transformations between different color spaces. It is important to recognize the distinction between color space and basis. A color space is defined by the visual subspace and corresponds to a specific individual, population of individuals, or device. We have already introduced several different color spaces in the preceding chapters based on different datasets, such as the 1931 CIE standard colorimetric observer, the 1964 CIE 10° colorimetric observer, and the Stiles and Burch 10° observer. For a given color space, we have different representations corresponding to different bases. As we will see, there can also be non-linear representations of a given color space, generally introduced to provide better perceptual uniformity. Transformations between color spaces are important because we often want to transform data acquired with a certain device such as a digital camera with its own color space to a standard color space. This always entails error since the equivalence classes for the two color spaces are different. Spectral densities which appear identical in one color space will appear different in the other, and vice versa.

5.1 LINEAR COLOR SPACE REPRESENTATIONS

Numerous color spaces and bases have been introduced in the literature for various purposes, but we just illustrate just one here, namely the 1931 CIE standard observer with ITU-R Rec. BT.709/ sRGB primaries.

There are many sets of red-green-blue (RGB) primaries that have been standardized over the years. A number of these are described in [40]. The most widely used set at this time is specified in the ITU-R Rec BT.709 standard for HDTV as well as the sRBG standard used in computing and defined in IEC 61966-2-1. These primaries are representative of the phosphors used in modern CRTs. They are considered to form a basis for the 1931 CIE space and are defined in terms of their 1931 XYZ chromaticities and the given reference white. This section will serve to illustrate many of the calculations done in colorimetry. These primaries can be denoted as $[\mathbf{R}_{709}]$, $[\mathbf{G}_{709}]$ and $[\mathbf{B}_{709}]$, and the basis is referred to as $\mathcal{R}709$. For this section, we use the simplified notation $[\mathbf{R}] = [\mathbf{R}_{709}]$, $[\mathbf{G}] = [\mathbf{G}_{709}]$, $[\mathbf{B}] = [\mathbf{B}_{709}]$. The primaries and reference white are specified in Table 5.1. Of course, the Z-chromaticities do not *need* to be provided in the table since the sum of the values in each column is 1. Slightly different values for the transformation matrices shown below can be found in the literature, generally due to slightly different values of the xy chomaticities of $[\mathbf{D}_{65}]$. I use here the official CIE values rounded to four decimal digits, and so I obtain the same matrices as found in IEC 61966-2-1.

Table 5.1: XYZ chromaticities of ITU-R Rec. 709 red, green and blue primaries and reference white D_{65}.

	Red	Green	Blue	White, D_{65}
x	0.640	0.300	0.150	0.3127
y	0.330	0.600	0.060	0.3290
z	0.030	0.100	0.790	0.3583

D_{65} is a reference white specified by the CIE and is meant to be typical of daylight. A representative power spectral density is shown in Fig. 3.10. In this section, we use the simplified notation $[\mathbf{D}] = [\mathbf{D}_{65}]$. Normalization is provided by setting the relative luminance of reference white to unity, $D_L = 1$. It follows that

$$D_X = \frac{d_X}{d_Y} = 0.9505$$
$$D_Y = 1$$
$$D_Z = \frac{d_Z}{d_Y} = 1.0891$$

To find the XYZ tristimulus values of the red, green and blue primaries, we need to find their relative luminances R_L, G_L and B_L. These are determined through the constraint $[\mathbf{R}] + [\mathbf{G}] + [\mathbf{B}] = [\mathbf{D}]$. This equation implies the matrix equation

$$\begin{bmatrix} R_X \\ R_Y \\ R_Z \end{bmatrix} + \begin{bmatrix} G_X \\ G_Y \\ G_Z \end{bmatrix} + \begin{bmatrix} B_X \\ B_Y \\ B_Z \end{bmatrix} = \begin{bmatrix} D_X \\ D_Y \\ D_Z \end{bmatrix}. \tag{5.1}$$

Using $R_X = R_L \frac{r_X}{r_Y}$ and so on in the above equation with the chromaticities from Table 5.1 gives

$$\begin{bmatrix} 1.\overline{93} & 0.5 & 2.5 \\ 1.0 & 1.0 & 1.0 \\ 0.\overline{09} & 0.1\overline{6} & 13.1\overline{6} \end{bmatrix} \begin{bmatrix} R_L \\ G_L \\ B_L \end{bmatrix} = \begin{bmatrix} 0.9505 \\ 1.0 \\ 1.0891 \end{bmatrix}. \tag{5.2}$$

Solving this matrix equation, we find

$$\begin{bmatrix} R_L \\ G_L \\ B_L \end{bmatrix} = \begin{bmatrix} 0.2126 \\ 0.7152 \\ 0.0722 \end{bmatrix} \tag{5.3}$$

and substituting into the expressions $R_X = R_L \frac{r_X}{r_Y}$ etc., we find

$$\begin{bmatrix} [\mathbf{R}] \\ [\mathbf{G}] \\ [\mathbf{B}] \end{bmatrix} = \underbrace{\begin{bmatrix} 0.4124 & 0.2126 & 0.0193 \\ 0.3576 & 0.7152 & 0.1192 \\ 0.1805 & 0.0722 & 0.9505 \end{bmatrix}}_{\mathbf{A}^T_{\mathcal{R}709\to\mathcal{XYZ}}} \begin{bmatrix} [\mathbf{X}] \\ [\mathbf{Y}] \\ [\mathbf{Z}] \end{bmatrix}. \tag{5.4}$$

To convert tristimulus values between the two sets of primaries, suppose that

$$[C] = C_R[\mathbf{R}] + C_G[\mathbf{G}] + C_B[\mathbf{B}]$$
$$= C_X[\mathbf{X}] + C_Y[\mathbf{Y}] + C_Z[\mathbf{Z}].$$

Then, referring to Table 3.1,

$$\begin{bmatrix} C_X \\ C_Y \\ C_Z \end{bmatrix} = \underbrace{\begin{bmatrix} 0.4124 & 0.3576 & 0.1805 \\ 0.2126 & 0.7152 & 0.0722 \\ 0.0193 & 0.1192 & 0.9505 \end{bmatrix}}_{\mathbf{A}_{\mathcal{R}709\to\mathcal{XYZ}}} \begin{bmatrix} C_R \\ C_G \\ C_B \end{bmatrix} \tag{5.5}$$

and

$$\begin{bmatrix} C_R \\ C_G \\ C_B \end{bmatrix} = \underbrace{\begin{bmatrix} 3.2406 & -1.5372 & -0.4986 \\ -0.9689 & 1.8758 & 0.0415 \\ 0.0557 & -0.2040 & 1.0570 \end{bmatrix}}_{\mathbf{A}^{-1}_{\mathcal{R}709\to\mathcal{XYZ}}} \begin{bmatrix} C_X \\ C_Y \\ C_Z \end{bmatrix}. \tag{5.6}$$

Note that the matrix of Eq. 5.6 is obtained by inverting the matrix of Eq. 5.5 as shown, i.e., rounded to four digits after the decimal, and rounding the result to four digits after the decimal.

The color matching functions, obtained by transforming the XYZ color matching functions using $\mathbf{A}^{-1}_{\mathcal{R}709\to\mathcal{XYZ}}$ are very similar to those shown in Fig. 3.4.

5.2   DIGITAL CAMERA COLOR SPACES

As mentioned several times, color spaces can also correspond to electronic sensors such as digital cameras. Most digital cameras are designed to acquire images for human viewing; this is the only type we are concerned with here (other types could be intra-red cameras, remote-sensing cameras, etc.) Such cameras usually have a three-dimensional color space, corresponding to three spectral sensitivities. However, some modern cameras have color spaces of more than three dimensions, which can lead to more accurate colorimetry.

For the purpose of illustration, two specific digital cameras are introduced here, using measured data from [22]: the Canon 10D and the Nikon D7. The data provided by the authors of [22] are shown in Fig. 5.1(a), (b). In each case, the three curves shown span the corresponding visual subspace

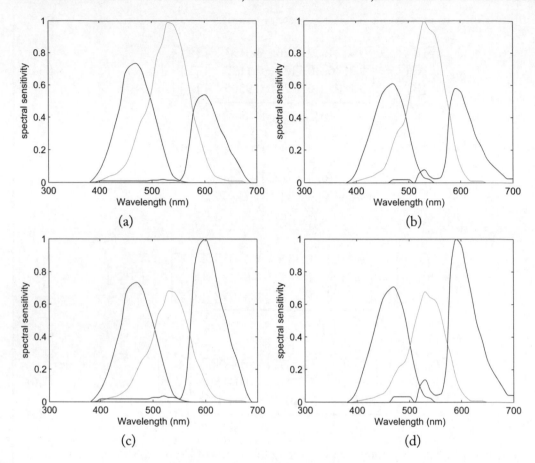

Figure 5.1: Spectral sensitivities of the Canon 10D and Nikon D70 digital cameras. (a) Canon 10D and (b) Nikon D70 show the raw spectral sensitivities from [22]. (c) Canon 10D and (d) Nikon D70 show the normalize spectral densities to match D65, as described in the text.

of the given camera and generate the corresponding color space. Each curve can have an arbitrary scaling without changing the color space. In keeping with the practice for human visual color spaces, the three curves for each camera have been scaled such that analysis of the reference white D65 with spectrum shown in Fig. 3.10 gives $k[1\ 1\ 1]^T$, with k arbitrarily selected so that the largest peak over the three curves is 1. These scaled spectral sensitivities are shown in Fig. 5.1(c), (d).

As with human visual color spaces, we can visualize coordinates obtained in a chromaticity diagram. Fig. 5.2 shows the spectrum locus in the two rg chromaticity diagrams, along with the convex hull. We see that unlike the case of human vision, parts of the spectrum locus lie within the convex hull.

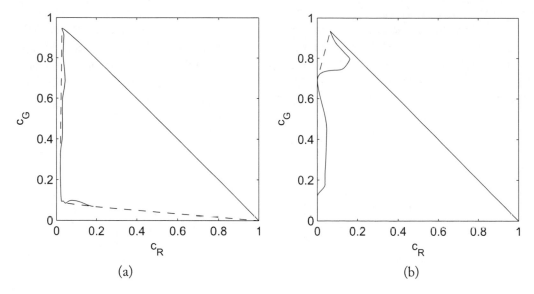

Figure 5.2: Chromaticity diagrams showing the spectrum locus and its convex hull. (a) Canon 10D. (b) Nikon D70.

5.3   NON-LINEAR COLOR COORDINATES

5.3.1   PERCEPTUALLY UNIFORM SPACES

Up to this point, there has been no mention of distance in color space. Colors are represented by their tristimulus values, but concepts of distance and orthogonality have not been defined. We can expect that two colors that have very similar tristimulus vectors should appear to be similar to each other, but we cannot apply any numerical measure to the difference. Indeed, it is well known that two pairs of colors with equal numerical difference in tristimulus vectors but residing in different regions of color space can have vastly different perceptual distance. There have been many attempts to equip color space with a metric, but due to nonlinear and adaptation properties of human vision, this is very difficult and beyond the scope of this lecture. The difficulty is very evident from Fig. 3.3, where the center portions of the upper diamonds appear to be quite different from each other, but correspond to the same point in color space. Thus, we must at least fix the same viewing conditions to talk about a simple distance metric in color space. Many more difficulties are discussed in [27]. Thus, I will just present a few basic facts about such distance measures for use in other aspects of this work.

The most commonly used color difference measures at this time are based on CIELAB. The CIELAB coordinates are obtained by applying a non-linear transformation to the XYZ tristimulus values, and they depend on a selected reference white. Either Euclidean distance or other more complex distance measures are then applied to the CIELAB coordinates to get the color difference measure. The goal is that equal values of the difference metric should correspond to equal perceptual difference in the colors. This is especially important for small differences as measuring large color differences (say between red and green) does not generally have any practical value. The CIELAB coordinates are denoted L^*, a^*, b^*. I will adapt the notation used for tristimulus values in this lecture to denote the CIELAB values as well. Thus, assume that $\mathbf{C}_{\mathcal{XYZ}} = [\, C_X \; C_Y \; C_Z \,]^T$ is the tristimulus vector for a color [C] in the $\mathcal{XYZ}$ basis, and that $\mathbf{W}_{\mathcal{XYZ}} = [\, W_X \; W_Y \; W_Z \,]^T$ is the selected reference white (e.g., $[\mathbf{D}_{65}]$). Define the non-linear real-valued function $f(x)$ on $[0,1]$,

$$f(x) = \begin{cases} \frac{1}{3}\left(\frac{29}{6}\right)^2 x + \frac{4}{29} & \text{if } 0 \leq x \leq \left(\frac{6}{29}\right)^3 \\ x^{1/3} & \text{if } \left(\frac{6}{29}\right)^3 \leq x \leq 1 \end{cases} \tag{5.7}$$

Note that $f(x)$ is one-to-one, monotonically increasing and continuous with continuous first derivative on $[0,1]$. For physically realizable colors, we assume $0 \leq C_i \leq W_i$, for $i \in \{X, Y, Z\}$ and CIELAB coordinates are defined for colors in the set $\{[0, W_X] \times [0, W_Y] \times [0, W_Z]\} \cap \mathcal{C}_R$.

Then the CIELAB coordinates are given by

$$\begin{aligned} C_{L^*} &= 116 f(C_Y/W_Y) - 16 \\ C_{a^*} &= 500(f(C_X/W_X) - f(C_Y/W_Y)) \\ C_{b^*} &= 200(f(C_Y/W_Y) - f(C_Z/W_Z)). \end{aligned} \tag{5.8}$$

We note that for any gray color $[\mathbf{G}_\alpha] = \alpha[\mathbf{W}]$ for $0 \leq \alpha \leq 1$, $G_{\alpha,a^*} = G_{\alpha,b^*} = 0$ and only G_{α,L^*} is non-zero, with values ranging from 0 to 100.

A measure of the difference between two colors $[\mathbf{C}_1]$ and $[\mathbf{C}_2]$ is

$$\Delta^*_{ab}([\mathbf{C}_1], [\mathbf{C}_2]) = \sqrt{(C_{1,L^*} - C_{2,L^*})^2 + (C_{1,a^*} - C_{2,a^*})^2 + (C_{1,b^*} - C_{2,b^*})^2}. \tag{5.9}$$

This formula is by no means perfect and more elaborate color difference formulas have been developed. See [58] for a recent discussion of efforts in this direction.

5.3.2   DEVICE-DEPENDENT COORDINATES

Perhaps the most widely used representation for digital color images, the device-dependent sRGB representation is a non-linear coordinate system historically related to the CRT display device. The light output of a CRT is related to the voltage applied approximately by a power law,

$$\text{intensity} = \text{voltage}^\gamma.$$

This applies to each of the RGB components of a display device, so that if the voltages applied are C'_R, C'_G and C'_B, the tristimulus values of the color emitted are $C_R = g(C'_R)$, $C_G = g(C'_G)$ and $C_B = g(C'_B)$, where $g(\cdot)$ is an invertible pointwise nonlinearity such as the power law mentioned above. Several such non-linearities have been standardized, such as the ones used in the sRGB and Rec. 709 standards mentioned previously. For example, the power-law non-linearity given in sRGB is

$$g(x) = \begin{cases} \frac{x}{12.92} & 0 \le x \le 0.04045 \\ \left(\frac{x+0.055}{1.055}\right)^{2.4} & 0.04045 \le x \le 1 \end{cases} \tag{5.10}$$

Note that the function has a linear segment near the origin. The function is monotonically increasing, continuous with continuous derivative on [0,1]. The inverse function is given by

$$g^{-1}(x) = \begin{cases} 12.92x & 0 \le x \le 0.0031308 \\ 1.055x^{0.41666} - 0.055 & 0.0031308 \le x \le 1. \end{cases} \tag{5.11}$$

Thus, if RGB tristimulus values are passed through $g^{-1}(\cdot)$, referred to as gamma correction, they can be applied directly to a CRT display to produce the correct tristimulus values. Today, as CRT displays are disappearing, other displays such as LCD displays apply the gamma function electronically for compatibility. Note that the non-linear transfer characteristics of displays will not in general be identical to the function of Eq. (5.10) with its linear portion near the origin, but they will closely approximate it to give good results. The gamma-corrected sRGB space is widely used, with most digital color images implicitly or explicitly assumed to be in this representation. Although this representation is not nearly as perceptually uniform as CIELAB, it is much more so than the linear tristimulus representation, and it is sufficient for many situations. Various transformations of the gamma-corrected RGB values are widely used (or have been used over the years) in video and digital photography. See Poynton's book [40] for an excellent treatment of these non-linear coordinate systems.

5.4   TRANSFORMATION BETWEEN COLOR SPACES

Up to this point, we have only seen the change of *representation* for colors belonging to a given color space, and projections to lower dimensional color spaces isomorphic to a subspace. However, we have also seen that there are many different color spaces, so there is often a need to transform colors in one space C_1 to elements of another C_2. A key example is where colors are measured in the color space of a camera or scanner and need to be converted to a human visual color space such as the 1931 CIE space, for eventual display and/or printing for human viewing. We seek a transformation $\mathcal{T} : C_1 \to C_2$ to carry out this mapping, and we should optimize the transformation for best performance according to a suitable criterion. In this formulation, the color spaces need not be of the same dimension, and the mapping $\mathcal{T}$ is not necessarily linear. There has been a great deal of work on this topic as it is required for the correct operation of all color cameras and scanners.

Assume that the color spaces C_1 and C_2 are defined by their respective visual subspaces $\mathcal{VS}_1$ and $\mathcal{VS}_2$. The respective projectors are $S_1 : A \to C_1$ and $S_2 : A \to C_2$. Then a particular spectral density $Q(\lambda) \in A$ is mapped to the two color spaces as

$$\begin{aligned} S_1(Q(\lambda)) &= [\mathbf{Q}^{(1)}] \in C_1, \\ S_2(Q(\lambda)) &= [\mathbf{Q}^{(2)}] \in C_2. \end{aligned} \tag{5.12}$$

In general, there is no mapping that will exactly map $[\mathbf{Q}^{(1)}]$ to $[\mathbf{Q}^{(2)}]$ for all $Q(\lambda) \in \mathcal{P}$. Thus, we seek a mapping $\mathcal{T}$ such that some error measure between $\mathcal{T}([\mathbf{Q}^{(1)}])$ and $[\mathbf{Q}^{(2)}]$ is minimized over a suitable ensemble of elements of $\mathcal{P}$. In this discussion, we ignore noise in the measurement process, which is of course important in practice and treated in some of the key literature (e.g., [48], [54]).

We will assume here that $\dim(\mathcal{C}_1) = M \geq 3$ and $\dim(\mathcal{C}_2) = 3$, and that $\mathcal{C}_2$ is in fact a human trichromatic visual color space. However, the discussion can easily be generalized to arbitrary $\dim(\mathcal{C}_2)$. If $\mathcal{VS}_2 \subset \mathcal{VS}_1$, the transformation *can* be carried out without error. This condition, often referred to as the Luther-Ives condition [46], implies that the color matching functions of human vision can be expressed as linear combinations of the camera spectral sensitivities. Let us fix bases for the two color spaces, say $\mathcal{D} = \{[\mathbf{R}_1], \ldots [\mathbf{R}_M]\}$ for $\mathcal{C}_1$ and $\mathcal{B} = \{[\mathbf{P}_1], [\mathbf{P}_2], [\mathbf{P}_3]\}$ for $\mathcal{C}_2$, with associated color matching functions $\{\bar{r}_1(\lambda), \ldots, \bar{r}_M(\lambda)\}$ and $\{\bar{p}_1(\lambda), \bar{p}_2(\lambda), \bar{p}_3(\lambda)\}$, respectively. Then, we can write

$$\bar{p}_i(\lambda) = \sum_{k=1}^{M} t_{ik}\bar{r}_k(\lambda), \quad i = 1, 2, 3. \tag{5.13}$$

Using the matrix form for Eq. 3.40, we can write

$$
\begin{aligned}
\mathbf{Q}_{\mathcal{B}}^{(2)} &= \int_{\mathcal{V}} Q(\lambda) \begin{bmatrix} \bar{p}_1(\lambda) \\ \bar{p}_2(\lambda) \\ \bar{p}_3(\lambda) \end{bmatrix} d\lambda \\
&= \int_{\mathcal{V}} Q(\lambda) \mathbf{T}_{\mathcal{D}\to\mathcal{B}} \begin{bmatrix} \bar{r}_1(\lambda) \\ \vdots \\ \bar{r}_M(\lambda) \end{bmatrix} d\lambda \\
&= \mathbf{T}_{\mathcal{D}\to\mathcal{B}} \int_{\mathcal{V}} Q(\lambda) \begin{bmatrix} \bar{r}_1(\lambda) \\ \vdots \\ \bar{r}_M(\lambda) \end{bmatrix} d\lambda,
\end{aligned}
\tag{5.14}
$$

where $\mathbf{T}_{\mathcal{D}\to\mathcal{B}}$ is the $3 \times M$ matrix with elements t_{ik} specified in Eq. (5.13). It follows immediately that

$$\mathbf{Q}_{\mathcal{B}}^{(2)} = \mathbf{T}_{\mathcal{D}\to\mathcal{B}}\mathbf{Q}_{\mathcal{D}}^{(1)}, \tag{5.15}$$

determining the desired transformation $\mathcal{T}$ with respect to the given bases. In this case, $\mathcal{T}$ is indeed a linear map.

In practice, the basis for $\mathcal{VS}_1$ is usually determined by the physical acquisition system, where the functions $\bar{r}_i(\lambda)$, $i = 1, \ldots, M$ represent the spectral sensitivities of M acquisition channels, for example red, green and blue filters in a digital camera. Furthermore, $\mathcal{VS}_2$ will never be precisely a subspace of $\mathcal{VS}_1$, although this is usually a design target for the acquisition filters. In this more general case, there is a (generally infinite) set of points in $\mathcal{C}_2$ that can correspond to a given point $[\mathbf{Q}^{(1)}]$ in $\mathcal{C}_1$. Let $[Q^{(1)}(\lambda)]_{\boxminus_1} = \mathcal{S}_1^{-1}([\mathbf{Q}^{(1)}])$ be the equivalence class of all spectral densities in $\mathcal{A}$ that map to $[\mathbf{Q}^{(1)}] \in \mathcal{C}_1$, where $Q^{(1)}(\lambda)$ is an arbitrary element of the class. This is a coset of the infinite dimensional subspace ker S_1, namely $Q^{(1)}(\lambda) + \ker S_1$. However, we are only concerned with physically realizable stimuli in this set, given by $S_1^{-1}([\mathbf{Q}^{(1)}]) \cap \mathcal{P}$. Since we will be using this set repeatedly, following Schmitt [44], we denote the *physical metamer set*

$$\mathcal{F}_1([\mathbf{Q}^{(1)}]) = S_1^{-1}([\mathbf{Q}^{(1)}]) \cap \mathcal{P}. \tag{5.16}$$

Thus we define the function $\mathcal{M}$ from $\mathcal{C}_1$ to the set of all subsets of $\mathcal{C}_2$ by

$$\mathcal{M} : [\mathbf{Q}^{(1)}] \mapsto S_2(\mathcal{F}_1([\mathbf{Q}^{(1)}])). \tag{5.17}$$

Any element of $\mathcal{M}([\mathbf{Q}^{(1)}])$ is a possible reconstruction of $[\mathbf{Q}^{(1)}]$ in $\mathcal{C}_2$. Numerous algorithms have been developed to select the *best* one according to various criteria.

A closely related and essentially equivalent problem has also been extensively studied (e.g., [34]). Assume that certain reflecting surfaces produce metameric stimuli when viewed under a given illuminant $W_1(\lambda)$. They may no longer be metameric when viewed under a different illuminant $W_2(\lambda)$. By multiplying a set of color matching functions by $W_1(\lambda)$ and $W_2(\lambda)$, respectively, we get the equivalent of two different color spaces. We do not explicitly pursue this variant of the problem here, although Schmitt [44] considered both versions.

Various terms have been used to describe the sets $\mathcal{M}([\mathbf{Q}^{(1)}])$ and their variants. The non-zero volume of these sets is due to the mismatch between the two visual subspaces, i.e., $\mathcal{VS}_2 \not\subset \mathcal{VS}_1$, and so we adopt the term *metamer mismatch set*, similar to [54] (who used metamer mismatch space) and [44] (who used metamer mismatch volume).

We now establish important fundamental characteristics of the metamer mismatch set. However, we first apply a necessary constraint to the color space $\mathcal{C}_1$, to avoid the problematic situation where non-zero elements of $\mathcal{P}$ would be mapped to 0. We require that $\ker S_1 \cap \mathcal{P} = 0$, as is the case for the human visual color space (Theorem 3.16). Any camera which does not satisfy this condition would not be useful for color reproduction.

Before studying the properties of the metamer mismatch set, we establish certain properties of set of physical metamers $\mathcal{F}_1([\mathbf{Q}^{(1)}])$ in $\mathcal{A}$. To do this, we use the L_1 metric in $\mathcal{A}$, as discussed in Section 2.4. The following result was stated loosely by Schmitt [44].

Theorem 5.1   *For any $[\mathbf{Q}^{(1)}] \in \mathcal{C}_1$, the set $\mathcal{F}_1([\mathbf{Q}^{(1)}])$ is a convex, bounded and closed subset of $\mathcal{P} \subset \mathcal{A}$.*

Proof. We have already established that $\mathcal{P}$ is convex in Section 2.3. To show that $S_1^{-1}([\mathbf{Q}^{(1)}])$ is convex, let $Q^{(1)}(\lambda) + K_a(\lambda)$ and $Q^{(1)}(\lambda) + K_b(\lambda)$ be two arbitrary elements of $S_1^{-1}([\mathbf{Q}^{(1)}])$, where $K_a(\lambda), K_b(\lambda) \in \ker S_1$. Then, $\alpha(Q^{(1)}(\lambda) + K_a(\lambda)) + (1 - \alpha)(Q^{(1)}(\lambda) + K_b(\lambda)) = Q^{(1)}(\lambda) + \alpha K_a(\alpha) + (1 - \alpha)K_b(\lambda) \in Q^{(1)}(\lambda) + \ker S_1$ since $\alpha K_a(\alpha) + (1 - \alpha)K_b(\lambda) \in \ker S_1$. The intersection of convex sets is convex, so $\mathcal{F}_1([\mathbf{Q}^{(1)}]) = S_1^{-1}([\mathbf{Q}^{(1)}]) \cap \mathcal{P}$ is convex.

To show that the set is closed and bounded, let $Q_a(\lambda)$ and $Q_b(\lambda)$ be two distinct elements of $\mathcal{F}_1([\mathbf{Q}^{(1)}])$. By convexity of this set, $Q_b(\lambda) + \alpha(Q_a(\lambda) - Q_b(\lambda)) \in \mathcal{F}_1([\mathbf{Q}^{(1)}])$ for $0 \le \alpha \le 1$. Now, $\pm(Q_a(\lambda) - Q_b(\lambda)) \in \ker \mathcal{S}_1$ but $\pm(Q_a(\lambda) - Q_b(\lambda)) \notin \mathcal{P}$ by the assumption that $\ker \mathcal{S}_1 \cap \mathcal{P} = 0$. It follows that $Q_a(\lambda) - Q_b(\lambda)$ must have both positive and negative values over $\mathcal{V}$.

Thus, as α increases beyond 1, at some value $Q_b(\lambda) + \alpha(Q_a(\lambda) - Q_b(\lambda))$ will just become 0 at some wavelength(s) where $Q_a(\lambda) - Q_b(\lambda)$ is negative, and then will be negative for α larger than that value. Specifically, let

$$\alpha_{\max} = \sup\{\alpha \mid Q_b(\lambda) + \alpha(Q_a(\lambda) - Q_b(\lambda)) \ge 0 \text{ for all } \lambda \in \mathcal{V}\}. \tag{5.18}$$

Then $Q_b(\lambda) + \alpha_{\max}(Q_a(\lambda) - Q_b(\lambda)) = 0$ for some $\lambda \in \mathcal{V}$ and $Q_b(\lambda) + \alpha(Q_a(\lambda) - Q_b(\lambda)) < 0$ for the same λ if $\alpha > \alpha_{\max}$. It follows that $Q_b(\lambda) + \alpha_{\max}(Q_a(\lambda) - Q_b(\lambda))$ is on the boundary of $\mathcal{F}_1([\mathbf{Q}^{(1)}])$. In the same way, as α decreases below 0, at some point $Q_b(\lambda) + \alpha(Q_a(\lambda) - Q_b(\lambda))$ will just become 0 at some wavelength(s) where $Q_a(\lambda) - Q_b(\lambda)$ is positive, and then it will be negative for α smaller than that value. We thus define

$$\alpha_{\min} = \inf\{\alpha \mid Q_b(\lambda) + \alpha(Q_a(\lambda) - Q_b(\lambda)) \ge 0 \text{ for all } \lambda \in \mathcal{V}\}. \tag{5.19}$$

Then $Q_b(\lambda) + \alpha_{\min}(Q_a(\lambda) - Q_b(\lambda))$ is also on the boundary of $\mathcal{F}_1([\mathbf{Q}^{(1)}])$, and $Q_b(\lambda) + \alpha(Q_a(\lambda) - Q_b(\lambda)) \in \mathcal{F}_1([\mathbf{Q}^{(1)}])$ for $\alpha_{\min} \le \alpha \le \alpha_{\max}$. The distance between these two boundary points is

$$(\alpha_{\max} - \alpha_{\min}) \int_\mathcal{V} |Q_a(\lambda) - Q_b(\lambda)| \, d\lambda < \infty \tag{5.20}$$

Since $(Q_a(\lambda)$ and $Q_b(\lambda))$ were arbitrary, we can thus conclude that $\sup\{d(Q_a, Q_b) \mid Q_a, Q_b \in \mathcal{F}_1([\mathbf{Q}^{(1)}])\} < \infty$ and thus $\mathcal{F}_1([\mathbf{Q}^{(1)}])$ is bounded. $\qquad\square$

Theorem 5.2   *For any finite* $[\mathbf{Q}^{(1)}] \in \mathcal{C}_1$, *where* $\ker \mathcal{S}_1 \cap \mathcal{P} = 0$, *the metamer mismatch set* $\mathcal{M}([\mathbf{Q}^{(1)}])$ *is a convex, bounded, closed subset of* $\mathcal{C}_2$.

Proof. Since $\mathcal{F}_1([\mathbf{Q}^{(1)}])$ is a convex, bounded, closed subset of $\mathcal{A}$, it follows that $\mathcal{S}_2(\mathcal{F}_1([\mathbf{Q}^{(1)}]))$ is a convex, bounded, closed subset of $\mathcal{C}_2$.

$\qquad\square$

We may ask the question: Is there a *linear* transformation that maps all physically realizable colors $[\mathbf{Q}^{(1)}] \in \mathcal{C}_1$ into the corresponding set $\mathcal{M}([\mathbf{Q}^{(1)}])$ in $\mathcal{C}_2$? If so, this would lead us to suspect that it may be sufficient to limit ourselves to linear transformations. The following result shows that the answer is affirmative for certain convex cones within $\mathcal{S}_1(\mathcal{P})$.

Theorem 5.3   *If* $\mathcal{C}_1$ *and* $\mathcal{C}_2$ *are two color spaces, there exists a linear transformation* $\mathcal{T}$ *such that* $\mathcal{T}[\mathbf{Q}^{(1)}] \in \mathcal{M}([\mathbf{Q}^{(1)}])$ *for all* $[\mathbf{Q}^{(1)}]$ *within a convex cone generated by* M *elements of* $\mathcal{C}_1$.

Proof. Let $R_1(\lambda), R_2(\lambda), \ldots, R_M(\lambda) \in \mathcal{P}$ be M physically realizable spectral densities such that $\mathcal{S}_1(R_1(\lambda)), \mathcal{S}_1(R_2(\lambda)), \ldots, \mathcal{S}_1(R_M(\lambda))$ are linearly independent in $\mathcal{C}_1$ and $\dim(\mathrm{span}(\mathcal{S}_2(R_1(\lambda)), \mathcal{S}_2(R_2(\lambda)), \ldots, \mathcal{S}_2(R_M(\lambda)))) = 3$. Typically, any arbitrarily selected $R_i(\lambda)$ will satisfy these conditions.

Without loss of generality, we take $\mathcal{D} = \{[\mathbf{R}_k] = \mathcal{S}_1(R_k(\lambda)), k = 1, \ldots, M\}$ as the basis for $\mathcal{C}_1$. We choose $\mathcal{T}$ to map $\mathcal{S}_1(R_k(\lambda))$ to $\mathcal{S}_2(R_k(\lambda))$ for $k = 1, \ldots, M$; these conditions fully specify the linear transformation. If we write

$$\mathcal{S}_2(R_k(\lambda)) = \sum_{i=1}^{3} t_{ik}[\mathbf{P}_i], \quad k = 1, \ldots, M, \tag{5.21}$$

it follows that

$$\mathcal{T}([\mathbf{R}_k]) = \sum_{i=1}^{3} t_{ik}[\mathbf{P}_i]. \tag{5.22}$$

Then, for any $[\mathbf{Q}^{(1)}] \in \mathcal{C}_1$, expressed as $[\mathbf{Q}^{(1)}] = \sum_{k=1}^{M} Q_k^{(1)}[\mathbf{R}_k]$,

$$[\widehat{\mathbf{Q}}^{(2)}] = \mathcal{T}([\mathbf{Q}^{(1)}]) = \sum_{k=1}^{M} Q_k^{(1)} \sum_{i=1}^{3} t_{ik}[\mathbf{P}_i]$$
$$= \sum_{i=1}^{3} \left(\sum_{k=1}^{M} t_{ik} Q_k^{(1)} \right) [\mathbf{P}_i], \tag{5.23}$$

i.e., $\widehat{\mathbf{Q}}_{\mathcal{B}}^{(2)} = \mathbf{T}_{\mathcal{D}\to\mathcal{B}} \mathbf{Q}_{\mathcal{D}}^{(1)}$.

We can guarantee that this linear transform yields a result within $\mathcal{M}([\mathbf{Q}^{(1)}])$ for all $[\mathbf{Q}^{(1)}]$ within the convex cone generated by the rays $\alpha[\mathbf{R}_i]$, $i = 1, \ldots, M$, i.e.,

$$\{\sum_{i=1}^{M} \alpha_i[\mathbf{R}_i] \mid \alpha_i \geq 0\}. \tag{5.24}$$

For any such $[\mathbf{Q}^{(1)}] = \sum_{i=1}^{M} Q_i^{(1)}[\mathbf{R}_i]$ with $Q_i^{(1)} \geq 0$, it follows that $\sum_{i=1}^{M} Q_i^{(1)} R_i(\lambda) \in \mathcal{S}_1^{-1}([\mathbf{Q}^{(1)}]) \cap \mathcal{P}$ and thus $\mathcal{T}([\mathbf{Q}^{(1)}]) \in \mathcal{M}([\mathbf{Q}^{(1)}])$.

$\square$

In fact, it has been generally found that linear transformations are effectively as good as any nonlinear transformations that have been found for many specific applications under suitable optimization criteria. Although the method in the proof of Theorem 5.3 can give a reasonable solution, the details of a particular application scenario must be taken into account to find the best transformation for that application.

CHAPTER 6

Signals and Systems Theory for Still and Time-Varying Color Images

This chapter presents the theory of color signals, namely still and time-varying color images. A color image is a function of two (spatial) or three (spatiotemporal) independent variables with values in a color space. We introduce the ideas of signals, linear shift-invariant systems and Fourier transforms, in the continuous domain and in the sampled domain.

6.1 CONTINUOUS-DOMAIN SYSTEMS FOR COLOR IMAGES

A continuous-space color image is a function that assigns an element of a given color space $\mathcal{C}$ to each spatial coordinate (x, y) within a given image window. This image will be denoted $[\mathbf{C}](x, y)$. Note that we use the standard symbols x and y to denote horizontal and vertical position, symbols that have already be used to denote chromaticities in the CIE XYZ color space. It should normally be clear from context which is meant. A time-varying color image would be of the form $[\mathbf{C}](x, y, t)$. We will use a vector notation for the independent variables, and write $[\mathbf{C}](\boldsymbol{x})$ for either still or time-varying images, where $\boldsymbol{x}$ denotes either (x, y) or (x, y, t) according to context. The corresponding domain is denoted $\mathbb{R}^D$, where $D = 2$ or $D = 3$. If we choose a basis $\mathcal{B} = \{[\mathbf{P}_1], [\mathbf{P}_2], [\mathbf{P}_3]\}$ for $\mathcal{C}$, then the color signal can be written

$$[\mathbf{C}](\boldsymbol{x}) = C_1(\boldsymbol{x})[\mathbf{P}_1] + C_2(\boldsymbol{x})[\mathbf{P}_2] + C_3(\boldsymbol{x})[\mathbf{P}_3] \qquad (6.1)$$

so that the color image is represented by three scalar images in this basis. It may be convenient to represent the color signal as a column matrix of scalar signals with respect to the given basis:

$$\mathbf{C}_{\mathcal{B}}(\boldsymbol{x}) = \begin{bmatrix} C_1(\boldsymbol{x}) \\ C_2(\boldsymbol{x}) \\ C_3(\boldsymbol{x}) \end{bmatrix}. \qquad (6.2)$$

Of course, the representation can be changed to any other basis for the color space using Eq. (3.33) at each point $\boldsymbol{x}$. We will consider three-dimensional color spaces, corresponding to trichromatic human vision, but the ideas can easily be adapted to any other dimension for the color space. A physically realizable color image will be one such that $[\mathbf{C}](\boldsymbol{x}) \in \mathcal{C}_R$ for all $\boldsymbol{x}$. Such an image would normally correspond to a non-negative spectral density $C(\boldsymbol{x}, \lambda)$, projected onto the given color space using the appropriate color matching functions, again independently at each point $\boldsymbol{x}$.

The set of color images forms a vector space $\mathcal{U}$ under the obvious definitions of pointwise addition and scalar multiplication at each x using the vector space operations of the color space $\mathcal{C}$, as long as the set is closed under these operations. As an example, we could have

$$\mathcal{U} = \{[\mathbf{C}](x) \mid |C_i(x)| < b \text{ for all } x;\ i = 1, 2, 3;\ b < \infty\}, \tag{6.3}$$

a space of bounded color images. The vector space $\mathcal{U}$ is of course infinite dimensional. Strictly speaking, $[\mathbf{C}](x)$ is the value of the signal at coordinate x and is an element of $\mathcal{C}$. However, following conventional usage and to avoid introduction of an additional symbol (e.g., $[\vec{\mathbf{C}}]$) for elements of $\mathcal{U}$, we will also use $[\mathbf{C}](x)$ to denote elements of $\mathcal{U}$ where it is clear from context what is meant.

A *system* is any operator $\mathcal{H} : \mathcal{U} \to \mathcal{U}$ (or more generally $\mathcal{H} : \mathcal{U}_1 \to \mathcal{U}_2$) that takes a color image as input and produces another color image as output:

$$[\mathbf{Q}](x) = \mathcal{H}([\mathbf{C}](x)). \tag{6.4}$$

When needed, we will write $(\mathcal{H}[\mathbf{C}])(x)$ to denote $\mathcal{H}([\mathbf{C}](x))$ at coordinate x. We are mainly concerned with two specific classes of systems, linear systems and memoryless systems, and perhaps combinations of these. A *linear system* is one that satisfies

$$\mathcal{H}(\alpha_1[\mathbf{C}_1](x) + \alpha_2[\mathbf{C}_2](x)) = \alpha_1\mathcal{H}([\mathbf{C}_1](x)) + \alpha_2\mathcal{H}([\mathbf{C}_2](x)). \tag{6.5}$$

A memoryless system is one for which

$$(\mathcal{H}[\mathbf{C}])(x)|_{x=s} \quad \text{depends only on} \quad [\mathbf{C}](x)|_{x=s} \tag{6.6}$$

for all s.

A simple linear system is the translation system

$$\mathcal{T}_d : (\mathcal{T}_d[\mathbf{C}])(x) = [\mathbf{C}](x - d). \tag{6.7}$$

A shift-invariant system is one that commutes with the shift system,

$$\mathcal{H}(\mathcal{T}_d[\mathbf{C}](x)) = \mathcal{T}_d(\mathcal{H}[\mathbf{C}](x)) \quad \text{for any } d \in \mathbb{R}^D. \tag{6.8}$$

We will mainly consider shift-invariant systems. A memoryless shift-invariant system is characterized by an arbitrary mapping from $\mathcal{C}$ into $\mathcal{C}$, which is applied independently at each coordinate x. A simple example of a memoryless shift-invariant system is given by

$$\mathcal{H}[\mathbf{C}](x) = \sqrt{C_1(x)}[\mathbf{P}_1] + \sqrt{C_2(x)}[\mathbf{P}_2] + \sqrt{C_3(x)}[\mathbf{P}_3]. \tag{6.9}$$

As in the usual scalar case, linear shift-invariant systems for color signals are characterized by their impulse response. The approach for scalar systems is standard [56], and it can easily be extended to the vector-valued signal by introduction of a basis $\mathcal{B}$ for $\mathcal{C}$. We can rewrite Eq. (6.1) as

$$[\mathbf{C}](x) = \sum_{i=1}^{3} \left(\int_{\mathbb{R}^D} C_i(s)\delta(x - s)\,ds \right) [\mathbf{P}_i]$$

$$= \sum_{i=1}^{3} \int_{\mathbb{R}^D} (C_i(s)\,ds)\,\delta(x - s)[\mathbf{P}_i]. \tag{6.10}$$

Taking this as the superposition of color signals of the form $\delta(\boldsymbol{x} - \boldsymbol{s})[\mathbf{P}_i]$ with weights $C_i(\boldsymbol{s})\,d\boldsymbol{s}$, we can apply linearity and shift invariance to obtain

$$
\begin{aligned}
[\mathbf{Q}](\boldsymbol{x}) = \mathcal{H}[\mathbf{C}](\boldsymbol{x}) &= \sum_{i=1}^{3} \int_{\mathbb{R}^D} (C_i(\boldsymbol{s})\,d\boldsymbol{s})\mathcal{H}(\delta(\boldsymbol{x} - \boldsymbol{s})[\mathbf{P}_i]) \\
&= \sum_{i=1}^{3} \int_{\mathbb{R}^D} (C_i(\boldsymbol{s})\,d\boldsymbol{s})\mathcal{T}_{\boldsymbol{s}}\mathcal{H}(\delta(\boldsymbol{x})[\mathbf{P}_i]).
\end{aligned}
\tag{6.11}
$$

Continuity of the linear operator $\mathcal{H}$ is assumed here to allow this interchange of operations. $\mathcal{H}(\delta(\boldsymbol{x})[\mathbf{P}_i])$ is the response to an impulse in the one-dimensional subspace spanned by $[\mathbf{P}_i]$; it will not, in general, be confined to $\mathrm{span}([\mathbf{P}_i])$. (If a linear operator $\mathcal{H} : \mathcal{U} \to \mathcal{U}$ is applied to elements of a subspace $\mathcal{U}_s$ of $\mathcal{U}$, the output will not be confined to $\mathcal{U}_s$; this will only occur in the special case that $\mathcal{U}_s$ is invariant under $\mathcal{H}$; see section 6.4 of [19].) Thus, we define

$$
\mathcal{H}(\delta(\boldsymbol{x})[\mathbf{P}_i]) = \sum_{k=1}^{3} h_{ki}(\boldsymbol{x})[\mathbf{P}_k].
\tag{6.12}
$$

With this definition, we obtain

$$
\begin{aligned}
[\mathbf{Q}](\boldsymbol{x}) &= \sum_{i=1}^{3}\sum_{k=1}^{3} \left(\int_{\mathbb{R}^D} C_i(\boldsymbol{s})h_{ki}(\boldsymbol{x} - \boldsymbol{s})\,d\boldsymbol{s} \right) [\mathbf{P}_k] \\
&= \sum_{k=1}^{3} \left(\sum_{i=1}^{3} (C_i * h_{ki})(\boldsymbol{x}) \right) [\mathbf{P}_k],
\end{aligned}
\tag{6.13}
$$

where $C_i * h_{ki}$ denotes the conventional multidimensional convolution of scalar signals. Note that convolution is commutative, $C_i * h_{ki} = h_{ki} * C_i$ [36]. In this way, we see that a linear shift-invariant system for color signals is characterized by *nine* scalar-valued impulse responses.

6.1.1   FREQUENCY RESPONSE AND FOURIER TRANSFORM

A sinusoidal color signal of frequency $\boldsymbol{u}$ is defined as

$$
[\mathbf{C}](\boldsymbol{x}) = [\mathbf{A}] \cos(2\pi \boldsymbol{u} \cdot \boldsymbol{x} + \phi).
\tag{6.14}
$$

As in scalar signals and systems, it is very convenient to introduce complex exponential sinusoidal signals. In order to do this, we must introduce the *complexification* of $\mathcal{C}$ [59]. Specifically, we define $\mathcal{C}_{\mathbb{C}}$ as the set of formal sums

$$
\mathcal{C}_{\mathbb{C}} = \{[\mathbf{C}_1] + j[\mathbf{C}_2] \mid [\mathbf{C}_1], [\mathbf{C}_2] \in \mathcal{C}\},
\tag{6.15}
$$

where j is the imaginary unit. Vector addition and scalar multiplication with elements of $\mathbb{C}$ are done in the obvious fashion. Note that $\dim(\mathcal{C}_{\mathbb{C}}) = \dim(\mathcal{C})$ and any basis for $\mathcal{C}$ is also a basis for $\mathcal{C}_{\mathbb{C}}$. We

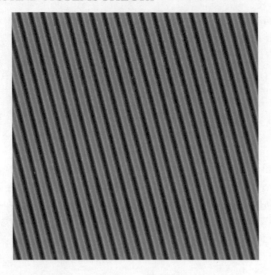

Figure 6.1:   Example of a real color sinusoidal signal.

can now define the complex exponential sinusoidal signals:

$$[\mathbf{C}](\boldsymbol{x}) = [\mathbf{A}]\exp(j2\pi\boldsymbol{u}\cdot\boldsymbol{x})$$
$$= \sum_{i=1}^{3} A_i \exp(j2\pi\boldsymbol{u}\cdot\boldsymbol{x})[\mathbf{P}_i] \tag{6.16}$$

expressed with respect to the basis $\mathcal{B}$, where $[\mathbf{A}]$ is arbitrary. Real sinusoidal color signals can be constructed as linear combinations of complex exponential sinusoidal signals. Fig. 6.1 shows the real sinusoidal signal

$$[\mathbf{C}](\boldsymbol{x}) = [\mathbf{A}_1] + [\mathbf{A}_2]\cos(2\pi\boldsymbol{u}\cdot\boldsymbol{x}) \tag{6.17}$$

where $\mathbf{A}_{1,sRGB} = \begin{bmatrix} 0.5 & 0.35 & 0.5 \end{bmatrix}^T$, $\mathbf{A}_{2,sRGB} = \begin{bmatrix} -0.5 & 0.35 & -0.5 \end{bmatrix}^T$ and $\boldsymbol{u} = (-20, 5)$ c/ph. This can be written as the sum of three complex exponentials,

$$[\mathbf{C}](\boldsymbol{x}) = [\mathbf{A}_1]\exp(j2\pi\mathbf{0}\cdot\boldsymbol{x}) + 0.5[\mathbf{A}_2]\exp(j2\pi\boldsymbol{u}\cdot\boldsymbol{x}) + 0.5[\mathbf{A}_2]\exp(j2\pi(-\boldsymbol{u})\cdot\boldsymbol{x}). \tag{6.18}$$

If we apply the complex sinusoidal signal $[\mathbf{C}](\boldsymbol{x})$ of Eq. (6.16) as input to an LSI system, then using Eq. (6.13) along with the commutativity of convolution, we obtain as output

$$[\mathbf{Q}](\boldsymbol{x}) = \sum_{k=1}^{3} \left(\sum_{i=1}^{3} \int_{\mathbb{R}^D} h_{ki}(\boldsymbol{s}) A_i \exp(j2\pi\boldsymbol{u}\cdot(\boldsymbol{x}-\boldsymbol{s}))\, d\boldsymbol{s} \right) [\mathbf{P}_k]$$
$$= \exp(j2\pi\boldsymbol{u}\cdot\boldsymbol{x}) \sum_{k=1}^{3} \left(\sum_{i=1}^{3} A_i \int_{\mathbb{R}^D} h_{ki}(\boldsymbol{s}) \exp(-j2\pi\boldsymbol{u}\cdot\boldsymbol{s})\, d\boldsymbol{s} \right) [\mathbf{P}_k]. \tag{6.19}$$

We identify

$$H_{ki}(u) = \int_{\mathbb{R}^D} h_{ki}(s) \exp(-j2\pi u \cdot s) \, ds \tag{6.20}$$

as the standard multidimensional Fourier transform of the scalar signal $h_{ki}(s)$. Thus

$$[\mathbf{Q}](x) = \left(\sum_{k=1}^{3} \left(\sum_{i=1}^{3} A_i H_{ki}(u) \right) [\mathbf{P}_k] \right) \exp(j2\pi u \cdot x). \tag{6.21}$$

Using the matrix notation

$$\mathbf{Q}_{\mathcal{B}}(x) = \begin{bmatrix} H_{11}(u) & H_{12}(u) & H_{13}(u) \\ H_{21}(u) & H_{22}(u) & H_{23}(u) \\ H_{31}(u) & H_{32}(u) & H_{33}(u) \end{bmatrix} \mathbf{A}_{\mathcal{B}} \exp(j2\pi u \cdot x)$$
$$= \mathbf{H}_{\mathcal{B}}(u)\mathbf{\Lambda}_{\mathcal{B}} \exp(j2\pi u \cdot x), \tag{6.22}$$

which defines the matrix $\mathbf{H}_{\mathcal{B}}(u)$; following our convention, $\mathbf{A}_{\mathcal{B}}$ is the column matrix of tristimulus values of $[\mathbf{A}]$ with respect to the basis $\mathcal{B}$.

Referring back to the general case of Eq. (6.13), the kth output component with respect to the basis $\mathcal{B}$ is

$$Q_k(x) = \sum_{i=1}^{3} (C_i * h_{ki})(x). \tag{6.23}$$

Applying the standard multidimensional convolution theorem [36],

$$\widehat{Q}_k(u) = \sum_{i=1}^{3} H_{ki}(u)\widehat{C}_i(u), \tag{6.24}$$

In matrix form, we can write

$$\widehat{\mathbf{Q}}_{\mathcal{B}}(u) = \mathbf{H}_{\mathcal{B}}(u)\widehat{\mathbf{C}}_{\mathcal{B}}(u) \tag{6.25}$$

where

$$\widehat{\mathbf{Q}}_{\mathcal{B}}(u) = \begin{bmatrix} \widehat{Q}_1(u) \\ \widehat{Q}_2(u) \\ \widehat{Q}_3(u) \end{bmatrix} \qquad \widehat{\mathbf{C}}_{\mathcal{B}}(u) = \begin{bmatrix} \widehat{C}_1(u) \\ \widehat{C}_2(u) \\ \widehat{C}_3(u) \end{bmatrix} \tag{6.26}$$

and $\widehat{C}_i(u)$ is the multidimensional Fourier transform of $C_i(x)$,

$$\widehat{C}_i(u) = \int_{\mathbb{R}^D} C_i(x) \exp(-j2\pi u \cdot x) \, dx \tag{6.27}$$

and similarly for $\widehat{Q}_i(u)$. We can consider $\widehat{\mathbf{C}}_{\mathcal{B}}(u)$ to be the *basis-dependent* form of the Fourier transform of the color signal $\mathbf{C}_{\mathcal{B}}(x)$. We could formally express the *basis-independent* version as

$$[\widehat{\mathbf{C}}](u) = \int_{\mathbb{R}^D} [\mathbf{C}](x) \exp(-j2\pi u \cdot x) \, dx \tag{6.28}$$

where we can interpret $[\widehat{\mathbf{C}}](\boldsymbol{u})$ as a mapping from the frequency domain $\mathbb{R}^D$ to $\mathcal{C}_{\mathbb{C}}$. In practice, we would introduce a basis to actually compute Eq. (6.28).

As already mentioned, if we change the basis of the color space, say from $\mathcal{B}$ to $\mathcal{B}'$, then the space-domain and the frequency-domain representations are transformed using the appropriate change-of-basis matrix as per Eq. (3.33):

$$\begin{aligned}\mathbf{C}_{\mathcal{B}'}(\boldsymbol{x}) &= \mathbf{A}_{\mathcal{B}\to\mathcal{B}'}\mathbf{C}_{\mathcal{B}}(\boldsymbol{x}), \\ \widehat{\mathbf{C}}_{\mathcal{B}'}(\boldsymbol{u}) &= \mathbf{A}_{\mathcal{B}\to\mathcal{B}'}\widehat{\mathbf{C}}_{\mathcal{B}}(\boldsymbol{u}).\end{aligned} \tag{6.29}$$

If we apply this to Eq. (6.25), we find

$$\widehat{\mathbf{Q}}_{\mathcal{B}'}(\boldsymbol{u}) = \mathbf{A}_{\mathcal{B}\to\mathcal{B}'}\mathbf{H}_{\mathcal{B}}(\boldsymbol{u})\mathbf{A}_{\mathcal{B}\to\mathcal{B}'}^{-1}\widehat{\mathbf{C}}_{\mathcal{B}'}(\boldsymbol{u}) \tag{6.30}$$

from which we conclude that

$$\mathbf{H}_{\mathcal{B}'}(\boldsymbol{u}) = \mathbf{A}_{\mathcal{B}\to\mathcal{B}'}\mathbf{H}_{\mathcal{B}}(\boldsymbol{u})\mathbf{A}_{\mathcal{B}\to\mathcal{B}'}^{-1}. \tag{6.31}$$

Thus, all transfer matrices for a given LSI color system with respect to different bases are equivalent and related by a similarity transformation as in Eq. (6.31).

Poirson and Wandell [38], [39] have found bases (for specific observers) such that the first stages of the human visual system can be approximated by a linear system with a diagonal transfer matrix in the given bases. Based on that work, Zhang and Wandell defined the spatial CIELAB (S-CIELAB) model [64]. This is intended to model the initial stages of the human visual system and was specifically developed in the context of an error metric for color images. I present it here as a good example of a continuous-space linear system for color images. In a specific basis, referred to as an opponent-colors basis, the transfer matrix of the linear system is diagonal. The linear system is followed by the conversion to nonlinear CIELAB coordinates. Assume that an image is represented in the CIE 1931 color space. Let us refer to the opponent-colors basis defined by Zhang and Wandell as $\mathcal{OZW} = \{[\mathbf{O}_1], [\mathbf{O}_2], [\mathbf{O}_3]\}$. This basis is specified by the transformation matrix

$$\mathbf{A}_{\mathcal{XYZ}\to\mathcal{OZW}} = \begin{bmatrix} 0.279 & 0.720 & -0.107 \\ -0.449 & 0.290 & -0.077 \\ 0.086 & -0.590 & 0.501 \end{bmatrix}. \tag{6.32}$$

In this basis, the transfer matrix $\mathbf{H}_{\mathcal{OZW}}(\boldsymbol{u})$ is diagonal. In the model, each term is circularly symmetric and formed as a weighted sum of Gaussian filters where the parameters of the filters were obtained empirically to best fit the Poirson and Wandell data. Reproducing the model of [64] in the present notation,

$$h_{ll}(x, y) = \sum_{i=1}^{I_l} w_{il}k_{il} \exp(-(x^2 + y^2)/\sigma_{il}^2), \quad l = 1, 2, 3. \tag{6.33}$$

The scale factors k_{il} are chosen so that the corresponding Gaussian filters have a DC gain (response at $(u, v) = (0, 0)$) of 1.0. The w_{il} are scaled such that the overall filter has a DC gain of unity, $H_{ll}(0, 0) = 1.0$.

Using the fact that the two-dimensional Fourier transform of the Gaussian impulse response is given by [36]

$$\exp(-(x^2 + y^2)/\sigma_{il}^2) \Leftrightarrow \pi \sigma_{il}^2 \exp(-\pi^2(u^2 + v^2)\sigma_{il}^2), \tag{6.34}$$

we see that the DC gain is 1.0 if $k_{il} = 1/(\pi \sigma_{il}^2)$. Thus, in the frequency domain, the filter responses are

$$H_{ll}(u, v) = \sum_{i=1}^{I_l} w_{il} \exp(-\pi^2(u^2 + v^2)\sigma_{il}^2), \tag{6.35}$$

and of course $H_{lm}(u, v) = 0$ for $l \neq m$. In any other basis, the transfer matrix is not diagonal, and it can be obtained using Eq. 6.31.

The empirical parameters from [64] are shown in Table 6.1, along with the above derived k_{il}. The weights are different than in [64] as they have been scaled to sum to 1.0, as is done in the widely available software that accompanies [64] (see [63]). Note that $I_1 = 3$, $I_2 = I_3 = 2$, and the unit of distance for x, y and σ_{il} is one degree of visual angle. The viewing distance is required to convert this to distance on the screen in a measure such as pixels or picture height. See [23] for an illustration of this conversion in the context of the present model. (Note that the σ parameters given in [23] are not the same as the ones in [63].)

Table 6.1: Parameters of the filters used in the S-CIELAB model.

component i	term l	w_{il}	σ_{il}	k_{il}
1	1	1.0033	0.0283	397.4
	2	0.1144	0.1330	17.99
	3	-0.1176	4.336	0.0169
2	1	0.6167	0.0392	207.1
	2	0.3833	0.4940	1.304
3	1	0.5681	0.0536	110.8
	2	0.4319	0.3860	2.136

6.2   DISCRETE-DOMAIN COLOR IMAGES

Although real-world physical color images such as the images on the surface of a display or on a sensor are continuous-domain signals, they must be converted to a discrete-domain (sampled) representation for digital processing, storage and transmission. We first discuss the discrete-domain representation and processing of color images, and then we address the sampling and reconstruction of continuous-domain signals.

Discrete-domain images will be defined on lattices. A lattice is a regular array of points in $\mathbb{R}^D$ that is shift invariant; if the lattice is translated such that some lattice point is shifted to coincide with another lattice point, then all the lattice points will line up. In this lecture, we use the formulation and notation of [6], [10]. A summary of the pertinent definitions and results concerning lattices is given in Appendix D.

We consider first the case where the color signal is defined at all points of a lattice $\Lambda \subset \mathbb{R}^D$. The thornier (but very widely used) situation when different components with respect to some basis are defined on different sampling structures will follow.

6.2.1   COLOR SIGNALS WITH ALL COMPONENTS ON THE SAME LATTICE

Again, a common notation can be used for both still and time-varying imagery using a vector notation for the independent variables. A color image is denoted $[\mathbf{C}](x)$, $x \in \Lambda$. The theory and development are very similar to that for continuous-domain images. We again assume that the collection of color images of interest belongs to a vector space that is referred to as a signal space, e.g.,

$$\mathcal{U}_d = \{[\mathbf{C}](x) \mid x \in \Lambda; \ |C_i(x)| < b, \ i = 1, 2, 3; \ b < \infty\}, \tag{6.36}$$

where $C_i(x)$ are tristimulus values with respect to any basis for $\mathcal{C}$. A system is any operator $\mathcal{H}:$ $\mathcal{U}_d \to \mathcal{U}_d$, and the definitions of linear and memoryless systems are unchanged from Eq. (6.5) and Eq. (6.6). Similarly, the translation system has the same definition as in Eq. (6.7) with the proviso that $d \in \Lambda$, and shift invariance is given by Eq. (6.8) with $d \in \Lambda$.

The characterization of a linear shift-invariant system follows the same path, but is, in fact, simpler since we do not need Dirac deltas. We define the *unit sample* function $\delta_\Lambda(x)$ on a lattice Λ by

$$\delta_\Lambda(x) = \begin{cases} 1 & x = 0 \\ 0 & x \in \Lambda \backslash 0, \end{cases} \tag{6.37}$$

where $\Lambda \backslash 0$ denotes all points of Λ except 0. An arbitrary color signal can be expressed

$$\begin{aligned} [\mathbf{C}](x) &= \sum_{s \in \Lambda} [\mathbf{C}](s)\delta_\Lambda(x - s) \\ &= \sum_{i=1}^{3} \sum_{s \in \Lambda} C_i(s)\delta_\Lambda(x - s)[\mathbf{P}_i], \end{aligned} \tag{6.38}$$

which is a superposition of color signals of the form $\delta_\Lambda(x - s)[\mathbf{P}_i]$ with weights $C_i(s)$. Applying linearity and shift invariance

$$\begin{aligned} [\mathbf{Q}](x) = \mathcal{H}[\mathbf{C}](x) &= \sum_{i=1}^{3} \sum_{s \in \Lambda} C_i(s)\mathcal{H}(\delta_\Lambda(x - s)[\mathbf{P}_i]) \\ &= \sum_{i=1}^{3} \sum_{s \in \Lambda} C_i(s)\mathcal{T}_s\mathcal{H}(\delta_\Lambda(x)[\mathbf{P}_i]). \end{aligned} \tag{6.39}$$

Again, $\mathcal{H}(\delta_\Lambda(\boldsymbol{x})[\mathbf{P}_i])$ is the response of the LSI system to a unit sample in the one-dimensional subspace of $\mathcal{C}$ spanned by $[\mathbf{P}_i]$, and it is a general color signal (not necessarily confined to that subspace) that can be expressed in terms of the basis as

$$\mathcal{H}(\delta_\Lambda(\boldsymbol{x})[\mathbf{P}_i]) = \sum_{k=1}^{3} h_{ki}(\boldsymbol{x})[\mathbf{P}_k]. \tag{6.40}$$

With this definition of the h_{ki}, we have

$$\begin{aligned}
[\mathbf{Q}](\boldsymbol{x}) &= \sum_{i=1}^{3}\sum_{k=1}^{3}\left(\sum_{\boldsymbol{s}\in\Lambda} C_i(\boldsymbol{s})h_{ki}(\boldsymbol{x}-\boldsymbol{s})\right)[\mathbf{P}_k] \\
&= \sum_{k=1}^{3}\left(\sum_{i=1}^{3}(C_i * h_{ki})(\boldsymbol{s})\right)[\mathbf{P}_k]
\end{aligned} \tag{6.41}$$

which is formally the same as Eq. (6.13), but now $C_i * h_{ki}$ denotes discrete-domain multidimensional convolution on the lattice Λ. Discrete convolution is also commutative. So once again, a linear shift-invariant discrete domain system for color signals is fully characterized by nine scalar-valued unit sample responses.

A complex exponential signal on a lattice Λ again has the same definition as in the continuous domain case, as given in Eq. (6.16). However, there is an important difference. In the continuous domain, each different frequency vector $\boldsymbol{u} \in \mathbb{R}^D$ corresponds to a different sinusoidal signal, with increasing frequency as $|\boldsymbol{u}|$ increases. In the discrete domain,

$$\exp(j2\pi\boldsymbol{u}\cdot\boldsymbol{x}) = \exp(j2\pi(\boldsymbol{u}+\boldsymbol{r})\cdot\boldsymbol{x}), \qquad \boldsymbol{r}\in\Lambda^* \tag{6.42}$$

where Λ^* is the reciprocal lattice (see Appendix D for details). Thus all frequencies in the set $\{\boldsymbol{u}+\boldsymbol{r}\mid\boldsymbol{r}\in\Lambda^*\}$ correspond to the same sinusoidal signal. It follows that we only need to consider frequencies in one unit cell of Λ^*.

If we apply a sinusoidal signal as input to a linear shift-invariant system, we obtain

$$\begin{aligned}
[\mathbf{Q}](\boldsymbol{x}) &= \sum_{k=1}^{3}\left(\sum_{i=1}^{3}\sum_{\boldsymbol{s}\in\Lambda} h_{ki}(\boldsymbol{s})A_i\exp(j2\pi\boldsymbol{u}\cdot(\boldsymbol{x}-\boldsymbol{s}))\right)[\mathbf{P}_k] \\
&= \exp(j2\pi\boldsymbol{u}\cdot\boldsymbol{x})\sum_{k=1}^{3}\left(\sum_{i=1}^{3}A_i\sum_{\boldsymbol{s}\in\Lambda} h_{ki}(\boldsymbol{s})\exp(-j2\pi\boldsymbol{u}\cdot\boldsymbol{s})\right)[\mathbf{P}_k].
\end{aligned} \tag{6.43}$$

We identify

$$H_{ki}(\boldsymbol{u}) = \sum_{\boldsymbol{s}\in\Lambda} h_{ki}(\boldsymbol{s})\exp(-j2\pi\boldsymbol{u}\cdot\boldsymbol{s}) \tag{6.44}$$

as the discrete-domain Fourier transform on the lattice Λ. Note that $H_{ki}(\boldsymbol{u}) = H_{ki}(\boldsymbol{u}+\boldsymbol{r})$ for all $\boldsymbol{r}\in\Lambda^*$, so that $H_{ki}(\boldsymbol{u})$ is periodic. We need only specify it on one period, which is a unit cell of the reciprocal lattice Λ^*. Thus

$$[\mathbf{Q}](x) = \left(\sum_{k=1}^{3} \sum_{i=1}^{3} A_i H_{ki}(u)[\mathbf{P}_k] \right) \exp(j2\pi u \cdot x) \tag{6.45}$$

and formally we can use Eq. (6.22).

These developments lead us to define the Fourier transform for a general color signal defined on a lattice Λ as

$$[\widehat{\mathbf{C}}](u) = \sum_{x \in \Lambda} [\mathbf{C}](x) \exp(-j2\pi u \cdot x) \tag{6.46}$$

which can be implemented using any desired basis. All the relations for change of basis on the Fourier transform and the transfer matrix given in Eq. (6.29)-(6.31) apply unchanged.

6.2.2   COLOR SIGNALS WITH DIFFERENT COMPONENTS ON DIFFERENT SAMPLING STRUCTURES

It is common practice to sample color signals with different components (according to some basis) defined on different sampling structures; in fact, it is almost universally applied at some point in the imaging chain. The most notable cases are: image acquisition using color filter arrays (CFA), such as the Bayer array where red, green and blue components are sampled on different structures; image displays using color pixel arrays such as LCD RGB displays; luma-chroma representations where chroma is downsampled with respect to luma. Although the latter application usually applies to non-linear color spaces, I will discuss the theory here mainly for linear color spaces.

Suppose that a continuous-domain signal is defined with respect to a specific basis $\mathcal{B} = \{[\mathbf{P}_1], [\mathbf{P}_2], [\mathbf{P}_3]\}$,

$$[\mathbf{C}_c](x) = C_{c1}(x)[\mathbf{P}_1] + C_{c2}(x)[\mathbf{P}_2] + C_{c3}(x)[\mathbf{P}_3], \quad x \in \mathbb{R}^D. \tag{6.47}$$

We assume that the three components $C_{ci}(x)$, $i = 1, 2, 3$ are represented on three sampling structures Ψ_i, which are all subsets of a common lattice Λ. Although the Ψ_i may not necessarily be lattices (if they do not contain the origin), they are generally shifted versions of a sublattice of Λ. Thus, the sampled components are $C_{di}(x)$, $x \in \Psi_i$. Each component can be treated separately using standard scalar signal processing; this is the conventional way to do things. However, it may be advantageous to take a holistic view as in previous sections. It is often problematic that not all components are available at a given sample location. However, we can define a color signal on the super-lattice Λ,

$$[\mathbf{C}_d](x) = \sum_{i=1}^{3} \gamma_i C_{di}(x)[\mathbf{P}_i], \quad x \in \Lambda, \tag{6.48}$$

where we set $C_{di}(x) = 0$ for $x \in \Lambda \setminus \Psi_i$. Thus, at any point of Λ, there could be anywhere from 0 to 3 non-zero components. The coefficients γ_i are required to maintain the same average ratios of the components and are inversely related to the relative sampling densities. An explicit formula will be given shortly.

The Fourier transform of $[\mathbf{C}_d](\boldsymbol{x})$ is given by

$$
\begin{aligned}
[\widehat{\mathbf{C}}_d](\boldsymbol{u}) &= \sum_{i=1}^{3} \gamma_i \left(\sum_{\boldsymbol{x} \in \Lambda} C_{di}(\boldsymbol{x}) \exp(-j 2\pi \boldsymbol{u} \cdot \boldsymbol{x}) \right) [\mathbf{P}_i] \\
&= \sum_{i=1}^{3} \gamma_i \left(\sum_{\boldsymbol{x} \in \Psi_i} C_{di}(\boldsymbol{x}) \exp(-j 2\pi \boldsymbol{u} \cdot \boldsymbol{x}) \right) [\mathbf{P}_i].
\end{aligned}
\tag{6.49}
$$

Although $[\widehat{\mathbf{C}}_d](\boldsymbol{u})$ is periodic, with one period being a unit cell of Λ^*, individual components $\widehat{C}_{di}(\boldsymbol{u})$ may have a smaller period, and thus several copies of the basic spectrum lie in one unit cell of Λ^*.

A common task and preoccupation is to relate the sampled signal C_d defined in Eq. (6.48) to one in which all components are defined at *every* point of the same lattice Λ. This can be studied by adapting the methods of [9] developed for color filter arrays to the present vector space representation. We assume that the sampling scheme is periodic and that the periodicity is given by a sublattice Γ of Λ. This is illustrated in Fig. 6.2 for a Bayer color filter array sampling scheme. The basic lattice Λ is a rectangular lattice with equal horizontal and vertical spacing X. To simplify notation, we take X as the unit of length called the pixel height (px), so that $X = 1$ px. Then Λ is the integer Cartesian lattice $\Lambda = \mathbb{Z}^2$. In the Bayer structure, one period of the sampling structure is replicated on points of the sublattice $\Gamma = (2\mathbb{Z})^2$. The number of sample points in one period is given by the index of Γ in Λ, denoted $K = (\Lambda : \Gamma)$, where $K = 4$ for the Bayer structure. The lattice Λ can be partitioned into a union of cosets of the sublattice Γ in Λ (this should be a familiar concept by now). For any $\boldsymbol{x} \in \Lambda$, $\boldsymbol{x} + \Gamma$ is a coset; cosets are either identical or disjoint and there are K distinct cosets. We choose one element from each coset as a *coset representative* denoted $\boldsymbol{b}_k, k = 1, \ldots, K$. It follows that $\Lambda = \cup_{k=1}^{K} (\boldsymbol{b}_k + \Gamma)$. For the Bayer structure of Fig. 6.2, a suitable set of coset representatives is $\boldsymbol{b}_1 = [0 \ 0]^T$, $\boldsymbol{b}_2 = [1 \ 0]^T$, $\boldsymbol{b}_3 = [0 \ 1]^T$, $\boldsymbol{b}_4 = [1 \ 1]^T$. For convenience, we arrange the coset representatives into a $2 \times K$ matrix $\mathbf{B} = [\boldsymbol{b}_1 \ \boldsymbol{b}_2 \ \cdots \ \boldsymbol{b}_K]$.

The key concept here is that each component C_{di} is defined on a selected subset of the set of K cosets. Specifically, the sampling structure Ψ_i of component i is given by

$$
\Psi_i = \bigcup_{k \in \mathcal{J}_i} (\boldsymbol{b}_k + \Gamma)
\tag{6.50}
$$

where the set $\mathcal{J}_i \subset \{1, 2, \ldots, K\}$ identifies which cosets form Ψ_i. For the Bayer structure, we have $\mathcal{J}_R = \{2\}, \mathcal{J}_G = \{1, 4\}$ and $\mathcal{J}_B = \{3\}$, giving $\Psi_R = (\boldsymbol{b}_2 + \Gamma), \Psi_G = (\boldsymbol{b}_1 + \Gamma) \cup (\boldsymbol{b}_4 + \Gamma)$ and $\Psi_B = (\boldsymbol{b}_3 + \Gamma)$. Although the indexing of cosets and the choice of coset representatives from each coset are arbitrary, we choose coset representatives near the origin, and we choose $\boldsymbol{b}_1 = 0$ by convention. The assignment of cosets to sampling structures can be summarized with the $K \times 3$ matrix

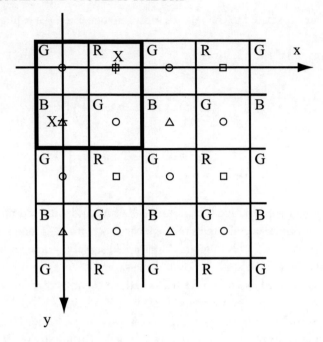

Figure 6.2: Bayer color filter array.

J defined by

$$J_{ki} = \begin{cases} 1 & k \in \mathcal{J}_i \\ 0 & \text{otherwise.} \end{cases} \qquad k = 1, \ldots, K; i = 1, 2, 3. \qquad (6.51)$$

With this formalism, we define the sampling functions

$$m_i(\boldsymbol{x}) = \begin{cases} 1 & \boldsymbol{x} \in \Psi_i \\ 0 & \boldsymbol{x} \in \Lambda \setminus \Psi_i. \end{cases} \qquad (6.52)$$

This can be used to relate the Fourier transform of $[\mathbf{C}_d](\boldsymbol{x})$ to the Fourier transform of a hypothetical signal $[\mathbf{C}](\boldsymbol{x})$ in which all components are defined on all of Λ. Specifically, we have

$$C_{di}(\boldsymbol{x}) = C_i(\boldsymbol{x})m_i(\boldsymbol{x}). \qquad (6.53)$$

Now each of the $m_i(\boldsymbol{x})$ is a periodic function on Λ, with $m_i(\boldsymbol{x} + \boldsymbol{y}) = m_i(\boldsymbol{x})$ for all $\boldsymbol{y} \in \Gamma$, and so it can be expressed using the discrete Fourier series. Since $\Gamma \subset \Lambda$, the inverse relation for reciprocal lattices is $\Lambda^* \subset \Gamma^*$ with $(\Gamma^* : \Lambda^*) = K$. We let $\{\boldsymbol{d}_1, \ldots, \boldsymbol{d}_K\}$ be a suitable set of coset representatives for Λ^* in Γ^*, summarized as before in the $2 \times K$ matrix $\mathbf{D} = [\, \boldsymbol{d}_1 \; \cdots \; \boldsymbol{d}_K \,]$. Again, we let $\boldsymbol{d}_1 = 0$. Recall that the reciprocal lattices are in the frequency domain so that the $\boldsymbol{d}_i$ are frequency vectors.

Continuing to use the formulation of [9], the Fourier series representation of the $m_i(x)$ is given by (see Appendix D)

$$m_i(x) = \sum_{k=1}^{K} M_{ki} \exp(j2\pi x \cdot d_k), \tag{6.54}$$

where

$$
\begin{aligned}
M_{ki} &= \frac{1}{K} \sum_{l=1}^{K} m_i(b_l) \exp(-j2\pi b_l \cdot d_k) \\
&= \frac{1}{K} \sum_{l \in \mathcal{J}_i} \exp(-j2\pi b_l \cdot d_k).
\end{aligned}
\tag{6.55}
$$

The previously defined matrix $\mathbf{J}$ can equivalently be defined by $J_{li} = m_i(b_l)$. We can then express the matrix $\mathbf{M}$ as

$$\mathbf{M} = \frac{1}{K}[\exp(-j2\pi \mathbf{D}^T \mathbf{B})]\mathbf{J} \tag{6.56}$$

where the exponential of the matrix is carried out term by term. We can now express the sampled color image as

$$
\begin{aligned}
[\mathbf{C}_d](x) &= \sum_{i=1}^{3} \gamma_i C_i(x) \sum_{k=1}^{K} M_{ki} \exp(j2\pi x \cdot d_k)[\mathbf{P}_i] \\
&= \sum_{k=1}^{K} \left(\sum_{i=1}^{3} \gamma_i M_{ki} C_i(x)[\mathbf{P}_i] \right) \exp(j2\pi x \cdot d_k) \\
&= \sum_{k=1}^{K} [\mathbf{F}_k](x) \exp(j2\pi x \cdot d_k),
\end{aligned}
\tag{6.57}
$$

where the new color signals $[\mathbf{F}_k](x)$ are defined as

$$[\mathbf{F}_k](x) = \sum_{i=1}^{3} \gamma_i M_{ki} C_i(x)[\mathbf{P}_i]. \tag{6.58}$$

Taking the Fourier transform, and using the modulation property,

$$[\widehat{\mathbf{C}}_d](u) = \sum_{k=1}^{K} [\widehat{\mathbf{F}}_k](u - d_k). \tag{6.59}$$

Thus, we can view the sampled signal as the sum of the baseband component $[\mathbf{F}_1](x)$ at zero frequency, and each of the other $[\mathbf{F}_k](x)$ shifted (modulated) to the non-zero frequencies d_k. Note that since $d_1 = 0$, then $M_{1i} = |\mathcal{J}_i|/K$. Thus, irrespective of the sampling scheme used, the baseband component $[\mathbf{F}_1](\boldsymbol{x}) = [\mathbf{C}](\boldsymbol{x})$ if

$$\gamma_i = \frac{1}{M_{1i}} = \frac{K}{|\mathcal{J}_i|}, \tag{6.60}$$

where $|\mathcal{J}_i|$ denotes the number of elements in the set $\mathcal{J}_i$. We will systematically use this choice for the γ_i.

Reference [9] develops this representation for scalar CFA signals for a number of different examples of color filter arrays. Each of these can be extended to the vector representation presented here. For this lecture, only the Bayer structure is examined in detail. The basis used for this structure is RGB, $[\mathbf{P}_1] = [\mathbf{R}]$, $[\mathbf{P}_2] = [\mathbf{G}]$, $[\mathbf{P}_3] = [\mathbf{B}]$. It does not matter which RGB basis is used (as long as it is known), and we can safely assume that it is the Rec. 709/sRGB basis of Section 5.1. For the Bayer structure, $\Lambda^* = \mathbb{Z}^2$ and $\Gamma^* = (\frac{1}{2}\mathbb{Z})^2$, and the following are suitable choices for the matrices defined above are:

$$\mathbf{B} = \begin{bmatrix} 0 & 1 & 0 & 1 \\ 0 & 0 & 1 & 1 \end{bmatrix}$$
$$\mathbf{D} = \begin{bmatrix} 0 & \frac{1}{2} & \frac{1}{2} & 0 \\ 0 & \frac{1}{2} & 0 & \frac{1}{2} \end{bmatrix}$$
$$\mathbf{J} = \begin{bmatrix} 0 & 1 & 0 \\ 1 & 0 & 0 \\ 0 & 0 & 1 \\ 0 & 1 & 0 \end{bmatrix} \tag{6.61}$$
$$\mathbf{M} = \frac{1}{4} \begin{bmatrix} 1 & 2 & 1 \\ -1 & 2 & -1 \\ -1 & 0 & 1 \\ 1 & 0 & -1 \end{bmatrix}$$

We can identify the baseband component as

$$[\mathbf{F}_1](\boldsymbol{x}) = \frac{1}{4}\gamma_1 C_1(\boldsymbol{x})[\mathbf{P}_1] + \frac{1}{2}\gamma_2 C_2(\boldsymbol{x})[\mathbf{P}_2] + \frac{1}{4}\gamma_3 C_3(\boldsymbol{x})[\mathbf{P}_3]. \tag{6.62}$$

The baseband component will be equal to the desired color signal $[\mathbf{C}](\boldsymbol{x})$ if $\gamma_1 = 4$, $\gamma_2 = 2$ and $\gamma_3 = 4$, in agreement with Eq. (6.60). With these values, the remaining three transformed signals are

$$[\mathbf{F}_2](\boldsymbol{x}) = -C_1(\boldsymbol{x})[\mathbf{P}_1] + C_2(\boldsymbol{x})[\mathbf{P}_2] - C_3(\boldsymbol{x})[\mathbf{P}_3]$$
$$[\mathbf{F}_3](\boldsymbol{x}) = -C_1(\boldsymbol{x})[\mathbf{P}_1] + C_3(\boldsymbol{x})[\mathbf{P}_3] \tag{6.63}$$
$$[\mathbf{F}_4](\boldsymbol{x}) = C_1(\boldsymbol{x})[\mathbf{P}_1] - C_3(\boldsymbol{x})[\mathbf{P}_3] = -[\mathbf{F}_3](\boldsymbol{x})$$

and in the frequency domain, we have

$$[\widehat{\mathbf{C}}_d](u, v) = [\widehat{\mathbf{F}}_1](u, v) + [\widehat{\mathbf{F}}_2](u - \frac{1}{2}, v - \frac{1}{2}) + [\widehat{\mathbf{F}}_3](u - \frac{1}{2}, v) - [\widehat{\mathbf{F}}_3](u, v - \frac{1}{2}). \tag{6.64}$$

Note that for convenience, we have used a different indexing of the frequency domain cosets than the one used in [9].

This representation suggests a frequency-domain approach to recover $[\mathbf{C}](\boldsymbol{x})$ (i.e., $[\mathbf{F}_1](\boldsymbol{x})$) from the subsampled signal $[\mathbf{C}_d](\boldsymbol{x})$, namely to isolate and remove the components at the frequencies $(\frac{1}{2}, 0)$, $(0, \frac{1}{2})$ and $(\frac{1}{2}, \frac{1}{2})$ with suitable filters. Using linear filters on the lattice Λ, the filters would be characterized by the 3×3 transfer matrices, as defined in Section 6.2.1. Doing separate processing on the $[\mathbf{P}_1]$, $[\mathbf{P}_2]$ and $[\mathbf{P}_3]$ (i.e., RGB) components would not work well. This simply amounts to separate interpolation of the subsampled RGB components, a method known to give relatively poor results [17]. An alternative approach to full vector processing is to perform a change of basis so that separate processing of the transformed coordinates is effective. We also note that the component $[\mathbf{F}_3](\boldsymbol{x})$ appears separately at frequencies $(\frac{1}{2}, 0)$ and $(0, \frac{1}{2})$, interacting differently with the neighboring components $[\mathbf{F}_1](\boldsymbol{x})$ and $[\mathbf{F}_2](\boldsymbol{x})$ located at the frequencies $(0, 0)$ and $(\frac{1}{2}, \frac{1}{2})$, respectively. This suggests the use of a locally adaptive filtering that weights more heavily the component suffering the least from crosstalk. This has been found to be a very effective approach in scalar demosaicking [7], [8].

Little work has been reported on such vector-based approaches and much investigation remains to be done. Just a few preliminary ideas are presented here. One interesting basis to consider for independent processing of components arises in the frequency-domain analysis of a CFA signal, obtained by multiplexing the red, green and blue samples defined on Ψ_1, Ψ_2 and Ψ_3 into a single scalar signal defined on $\Lambda = \Psi_1 \cup \Psi_2 \cup \Psi_3$,

$$C_{\text{CFA}}(\boldsymbol{x}) = C_1(\boldsymbol{x})m_1(\boldsymbol{x}) + C_2(\boldsymbol{x})m_2(\boldsymbol{x}) + C_3(\boldsymbol{x})m_3(\boldsymbol{x}). \tag{6.65}$$

Using the Fourier series representation of the $m_i(\boldsymbol{x})$ of Eq. (6.54) with $K = 4$, we find

$$C_{\text{CFA}}(\boldsymbol{x}) = \sum_{k=1}^{4} \left(\sum_{i=1}^{3} M_{ki} C_i(\boldsymbol{x}) \right) \exp(j2\pi \boldsymbol{x} \cdot \boldsymbol{d}_k), \tag{6.66}$$

where $\mathbf{M}$ is given for the Bayer pattern in Eq. 6.61. Since the last row of $\mathbf{M}$ is the negative of the third row, we define only three new signals

$$\begin{bmatrix} C_L(\boldsymbol{x}) \\ C_{CH1}(\boldsymbol{x}) \\ C_{CH2}(\boldsymbol{x}) \end{bmatrix} = \underbrace{\begin{bmatrix} \frac{1}{4} & \frac{1}{2} & \frac{1}{4} \\ -\frac{1}{4} & \frac{1}{2} & -\frac{1}{4} \\ -\frac{1}{4} & 0 & \frac{1}{4} \end{bmatrix}}_{\mathbf{A}_{\mathcal{RGB} \to \mathcal{LCC}}} \begin{bmatrix} C_1(\boldsymbol{x}) \\ C_2(\boldsymbol{x}) \\ C_3(\boldsymbol{x}) \end{bmatrix}. \tag{6.67}$$

With this definition, we obtain

$$C_{\text{CFA}}(\boldsymbol{x}) = C_L(\boldsymbol{x}) + C_{CH1}(\boldsymbol{x}) \exp(j2\pi \boldsymbol{x} \cdot \boldsymbol{d}_2) + C_{CH2}(\boldsymbol{x})(\exp(j2\pi \boldsymbol{x} \cdot \boldsymbol{d}_3) - \exp(j2\pi \boldsymbol{x} \cdot \boldsymbol{d}_4)) \tag{6.68}$$

with the corresponding frequency domain representation

$$\widehat{C}_{\text{CFA}}(u, v) = \widehat{C}_L(u, v) + \widehat{C}_{CH1}(u - \frac{1}{2}, v - \frac{1}{2}) + \widehat{C}_{CH2}(u - \frac{1}{2}, v) - \widehat{C}_{CH2}(u, v - \frac{1}{2}). \tag{6.69}$$

In this process, three new signals have been defined in Eq. (6.67), which are a linear transformation of the original tristimulus values. Thus, these new values can be considered to be tristimulus values with respect to a new basis.

As per Table 3.1, this defines the new primaries forming the basis $\mathcal{LCC} = \{[\mathbf{L}], [\mathbf{CH}_1], [\mathbf{CH}_2]\}$,

$$
\begin{bmatrix} [\mathbf{L}] \\ [\mathbf{CH}_1] \\ [\mathbf{CH}_2] \end{bmatrix} = \underbrace{\begin{bmatrix} 1 & 1 & 1 \\ -1 & 1 & -1 \\ -2 & 0 & 2 \end{bmatrix}}_{\mathbf{A}_{\mathcal{RGB}\to\mathcal{LCC}}^{-T}} \begin{bmatrix} [\mathbf{P}_1] \\ [\mathbf{P}_2] \\ [\mathbf{P}_3] \end{bmatrix}. \tag{6.70}
$$

We observe that the basis vector $[\mathbf{L}] = [\mathbf{P}_1] + [\mathbf{P}_2] + [\mathbf{P}_3]$ is in fact reference white, and that for any gray-scale image in which $C_1(\boldsymbol{x}) = C_2(\boldsymbol{x}) = C_3(\boldsymbol{x})$, the two components $C_{CH1}(\boldsymbol{x})$ and $C_{CH2}(\boldsymbol{x})$ are zero.

Applying the transformation of Eq. (6.67) to the color signals $[\mathbf{F}_1](\boldsymbol{x})$, $[\mathbf{F}_2](\boldsymbol{x})$ and $[\mathbf{F}_3](\boldsymbol{x})$ defined in Eq. (6.62) and Eq. (6.63), we obtain

$$
\begin{aligned}
\mathbf{F}_{1,\mathcal{LCC}}(\boldsymbol{x}) &= \begin{bmatrix} \frac{1}{4}C_1(\boldsymbol{x}) + \frac{1}{2}C_2(\boldsymbol{x}) + \frac{1}{4}C_3(\boldsymbol{x}) \\ -\frac{1}{4}C_1(\boldsymbol{x}) + \frac{1}{2}C_2(\boldsymbol{x}) - \frac{1}{4}C_3(\boldsymbol{x}) \\ -\frac{1}{4}C_1(\boldsymbol{x}) + \frac{1}{4}C_3(\boldsymbol{x}) \end{bmatrix} = \begin{bmatrix} C_L(\boldsymbol{x}) \\ C_{CH1}(\boldsymbol{x}) \\ C_{CH2}(\boldsymbol{x}) \end{bmatrix} \\
\mathbf{F}_{2,\mathcal{LCC}}(\boldsymbol{x}) &= \begin{bmatrix} -\frac{1}{4}C_1(\boldsymbol{x}) + \frac{1}{2}C_2(\boldsymbol{x}) - \frac{1}{4}C_3(\boldsymbol{x}) \\ \frac{1}{4}C_1(\boldsymbol{x}) + \frac{1}{2}C_2(\boldsymbol{x}) + \frac{1}{4}C_3(\boldsymbol{x}) \\ \frac{1}{4}C_1(\boldsymbol{x}) - \frac{1}{4}C_3(\boldsymbol{x}) \end{bmatrix} = \begin{bmatrix} C_{CH1}(\boldsymbol{x}) \\ C_L(\boldsymbol{x}) \\ -C_{CH2}(\boldsymbol{x}) \end{bmatrix} \\
\mathbf{F}_{3,\mathcal{LCC}}(\boldsymbol{x}) &= \begin{bmatrix} -\frac{1}{4}C_1(\boldsymbol{x}) + \frac{1}{4}C_3(\boldsymbol{x}) \\ \frac{1}{4}C_1(\boldsymbol{x}) - \frac{1}{4}C_3(\boldsymbol{x}) \\ \frac{1}{4}C_1(\boldsymbol{x}) + \frac{1}{4}C_3(\boldsymbol{x}) \end{bmatrix} = \begin{bmatrix} C_{CH2}(\boldsymbol{x}) \\ -C_{CH2}(\boldsymbol{x}) \\ \frac{1}{2}(C_L(\boldsymbol{x}) - C_{CH1}(\boldsymbol{x})) \end{bmatrix} = -\mathbf{F}_{4,\mathcal{LCC}}(\boldsymbol{x})
\end{aligned} \tag{6.71}
$$

In this basis, $[\mathbf{C}_d](\boldsymbol{x})$ in Eq. (6.57) can be written out explicitly as

$$
\mathbf{C}_{d,\mathcal{LCC}}(\boldsymbol{x}) = \begin{bmatrix} C_L(\boldsymbol{x}) + C_{CH1}(\boldsymbol{x})\exp(j2\pi\boldsymbol{x}\cdot\boldsymbol{d}_2) + C_{CH2}(\boldsymbol{x}) \\ \quad(\exp(j2\pi\boldsymbol{x}\cdot\boldsymbol{d}_3) - \exp(j2\pi\boldsymbol{x}\cdot\boldsymbol{d}_4)) \\ C_{CH1}(\boldsymbol{x}) + C_L(\boldsymbol{x})\exp(j2\pi\boldsymbol{x}\cdot\boldsymbol{d}_2) - C_{CH2}(\boldsymbol{x}) \\ \quad(\exp(j2\pi\boldsymbol{x}\cdot\boldsymbol{d}_3) - \exp(j2\pi\boldsymbol{x}\cdot\boldsymbol{d}_4)) \\ C_{CH2}(\boldsymbol{x}) - C_{CH2}(\boldsymbol{x})\exp(j2\pi\boldsymbol{x}\cdot\boldsymbol{d}_2) + \frac{1}{2}(C_L(\boldsymbol{x}) - C_{CH1}(\boldsymbol{x})) \\ \quad(\exp(j2\pi\boldsymbol{x}\cdot\boldsymbol{d}_3) - \exp(j2\pi\boldsymbol{x}\cdot\boldsymbol{d}_4)) \end{bmatrix} \tag{6.72}
$$

In this basis, the $[\mathbf{L}]$ component is the conventional CFA signal. Given the definition of the $\boldsymbol{d}_i$ for the Bayer structure, it is straightforward to show that $C_{d,CH1}(\boldsymbol{x}) = C_{d,L}(\boldsymbol{x})\exp(2\pi(\boldsymbol{x}\cdot\boldsymbol{d}_2))$, and so the $[\mathbf{CH1}]$ component provides the same information as the $[\mathbf{L}]$ component. However, the $[\mathbf{CH2}]$ component depends only on $C_1(\boldsymbol{x})$ and $C_3(\boldsymbol{x})$ and can perhaps provide additional information about the original components and thereby give improved demosaicking performance compared to methods using only the $[\mathbf{L}]$ component (i.e., the conventional CFA signal). This remains to be seen.

6.3   ANALYSIS OF COLOR MOSAIC DISPLAYS

Most color image display devices, including CRT, LCD, plasma, etc., are based on a mosaic of color elements. A recent model and analysis for such displays has been presented by Farrell et al. [13]. In such models, it is assumed that the display screen is partitioned into non-overlapping regions. One of a finite number of base colors is displayed in each region; there are usually three base colors (red, green and blue), but recent work has considered more than three to expand the gamut of displayable colors (e.g., [29]). Using more than three base colors can allow a larger portion of the cone of physically realizable colors to be displayable by additive synthesis on a given device, as discussed in Section 3.10.2. We adapt our preceding development to such displays to describe the continuous-space color image on the screen. For the purpose of this lecture, I will assume three base colors (red, green and blue, denoted P_1, P_2, P_3), but the development can be extended to more than three base colors in a straightforward fashion. Assume that the three base colors at full intensity emit light with spectral density $P_i(\lambda)$, $i = 1, 2, 3$, with typical examples shown in Fig. 2.1.

According to the linear additive model, the displayed continuous-space image (expressed as spectral radiant exitance) is given by

$$C_o(\boldsymbol{x}, \lambda) = \sum_{i=1}^{3} \sum_{\boldsymbol{s} \in \Psi_i} C_{di}(\boldsymbol{s}) a(\boldsymbol{x} - \boldsymbol{s}) P_i(\lambda). \tag{6.73}$$

We assume that a specific color space is used to represent the displayed color image, usually the 1931 CIE standard observer. In this case, the spectral densities $P_i(\lambda)$ are projected to the color space in the usual fashion to obtain $[\mathbf{P}_i]$, $i = 1, 2, 3$. Then, the displayed color signal is given by

$$[\mathbf{C}_o](\boldsymbol{x}) = \sum_{i=1}^{3} \sum_{\boldsymbol{s} \in \Psi_i} C_{di}(\boldsymbol{s}) a(\boldsymbol{x} - \boldsymbol{s})[\mathbf{P}_i]. \tag{6.74}$$

In this model, we assume that the set of display elements lie on a lattice Λ that is partitioned as the union of three subsets, $\Lambda = \cup \Psi_i$, where the Ψ_i are cosets of a sublattice Γ in Λ. The continuous space function $a(\boldsymbol{x})$ is the display aperture. In typical mosaic displays such as an LCD display, shifted versions of $a(\boldsymbol{x})$ on Λ do not overlap, i.e., $a(\boldsymbol{x} - \boldsymbol{s}_1) \cdot a(\boldsymbol{x} - \boldsymbol{s}_2) = 0$ for $\boldsymbol{s}_1, \boldsymbol{s}_2 \in \Lambda$, $\boldsymbol{s}_1 \neq \boldsymbol{s}_2$. A common example of display mosaic is illustrated in Fig. 6.3(a), corresponding to the rectangular display aperture function in Fig. 6.3(b). A popular alternative aperture is the chevron of Fig. 6.3(c), which can tile the display screen according to the same arrangement as Fig. 6.3(a). See [13] for close-up photographs of display mosaics of each type.

We can directly compute the Fourier transform of the displayed signal of Eq. (6.74),

$$[\widehat{\mathbf{C}}_o](\boldsymbol{u}) = \sum_{i=1}^{3} \sum_{\boldsymbol{s} \in \Psi_i} C_{di}(\boldsymbol{s}) \int_{\mathbb{R}^2} a(\boldsymbol{x} - \boldsymbol{s}) \exp(-j 2\pi \boldsymbol{u} \cdot \boldsymbol{x}) \, d\boldsymbol{x}[\mathbf{P}_i]. \tag{6.75}$$

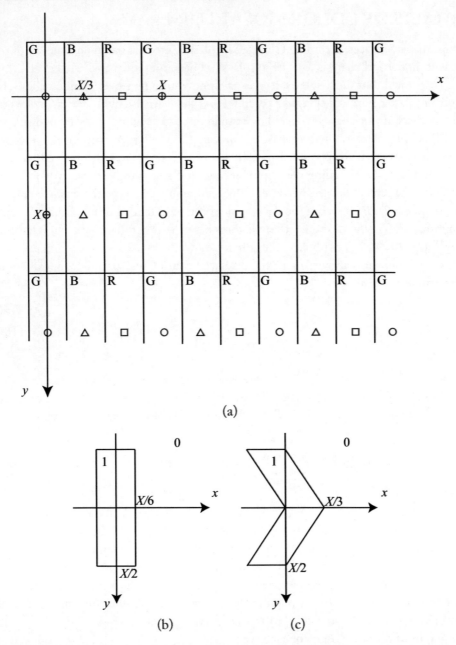

Figure 6.3: (a) Example of a typical display mosaic with a rectangular display aperture. (b) Rectangular aperture function. (c) Chevron display aperture.

Using the shift property of the continuous space Fourier transform,

$$\int_{\mathbb{R}^2} a(\boldsymbol{x} - \boldsymbol{s}) \exp(-j2\pi \boldsymbol{u} \cdot \boldsymbol{x}) \, d\boldsymbol{x} = \hat{a}(\boldsymbol{u}) \exp(-j2\pi \boldsymbol{u} \cdot \boldsymbol{x}) \tag{6.76}$$

so that

$$[\widehat{\mathbf{C}}_o](\boldsymbol{u}) = \hat{a}(\boldsymbol{u}) \sum_{i=1}^{3} \left(\sum_{\boldsymbol{s} \in \Psi_i} C_{di}(\boldsymbol{s}) \exp(-j2\pi \boldsymbol{u} \cdot \boldsymbol{x}) \right) [\mathbf{P}_i]. \tag{6.77}$$

We see that the latter term is essentially the same as Eq. (6.49) with $\gamma_i = 1$, and thus the analysis of the preceding section can be applied directly. Then, this periodic Fourier transform is multiplied by the aperiodic $\hat{a}(\boldsymbol{u})$ to get the Fourier transform of the displayed signal. This can be coupled with a model of the human visual system that accounts for the frequency response of the color channels, such as the S-CIELAB model of Zhang and Wandell [64] to optimize the display mosaic, as well as any processing to choose the $C_{di}(\boldsymbol{x})$ from the stored color image to be displayed (see Xu et al. [62] for an example).

To illustrate these ideas in the following, I identify all the parameters and matrices associated with the display mosaic of Fig. 6.3(a) in order to determine the Fourier transform in Eq. 6.77. By inspection of Fig. 6.3(a), we identify the following, using the notation of the preceding section.

$$\Gamma = \mathrm{LAT}\left(\begin{bmatrix} X & 0 \\ 0 & X \end{bmatrix} \right) \qquad \Lambda = \left(\begin{bmatrix} \frac{X}{3} & 0 \\ 0 & X \end{bmatrix} \right)$$

$$\Gamma^* = \mathrm{LAT}\left(\begin{bmatrix} \frac{1}{X} & 0 \\ 0 & \frac{1}{X} \end{bmatrix} \right) \qquad \Lambda^* = \mathrm{LAT}\left(\begin{bmatrix} \frac{3}{X} & 0 \\ 0 & \frac{1}{X} \end{bmatrix} \right)$$

$$K = 3$$

$$\mathbf{B} = \begin{bmatrix} 0 & \frac{X}{3} & \frac{2X}{3} \\ 0 & 0 & 0 \end{bmatrix} \qquad \mathbf{D} = \begin{bmatrix} 0 & \frac{1}{X} & -\frac{1}{X} \\ 0 & 0 & 0 \end{bmatrix}$$

$$\mathbf{J} = \begin{bmatrix} 1 & 0 \\ 0 & 1 \end{bmatrix}$$

$$\mathbf{M} = \frac{1}{3} \begin{bmatrix} 1 & 1 & 1 \\ 1 & e^{-j2\pi/3} & e^{j2\pi/3} \\ 1 & e^{j2\pi/3} & e^{-j2\pi/3} \end{bmatrix}$$

Finally, the three new signals defined are

$$[\mathbf{F}_1](\boldsymbol{x}) = \frac{1}{3}(C_{d1}(\boldsymbol{x})[\mathbf{P}_1] + C_{d2}(\boldsymbol{x})[\mathbf{P}_2] + C_{d3}(\boldsymbol{x})[\mathbf{P}_3])$$

$$[\mathbf{F}_2](\boldsymbol{x}) = \frac{1}{3}(C_{d1}(\boldsymbol{x})[\mathbf{P}_1] + e^{-j2\pi/3}C_{d2}(\boldsymbol{x})[\mathbf{P}_2] + e^{j2\pi/3}C_{d3}(\boldsymbol{x})[\mathbf{P}_3])$$

$$[\mathbf{F}_3](\boldsymbol{x}) = \frac{1}{3}(C_{d1}(\boldsymbol{x})[\mathbf{P}_1] + e^{j2\pi/3}C_{d2}(\boldsymbol{x})[\mathbf{P}_2] + e^{-j2\pi/3}C_{d3}(\boldsymbol{x})[\mathbf{P}_3])$$

and the desired Fourier transform is

$$[\widehat{\mathbf{C}}_o](u, v) = \hat{a}(u, v)([\widehat{\mathbf{F}}_1](u, v) + [\widehat{\mathbf{F}}_2](u - \frac{1}{X}, v) + [\widehat{\mathbf{F}}_3](u + \frac{1}{X}, v) \tag{6.78}$$

Note that if $C_{di}(x)$ is equal to the desired signal on Ψ_i, the baseband component is essentially the desired color signal, scaled by a factor of one third, since two out of three samples for each component are zero. The components at $\frac{1}{X}$ and $-\frac{1}{X}$ are due to the display mosaic. Higher spectral repeats are attenuated by a combination of the display aperture $\hat{a}(u, v)$ and the human visual system. The display aperture Fourier transform is easily found using standard methods, either analytically or using an approximation with the discrete Fourier transform. The overall perceptual effect of the display process can be modeled with the S-CIELAB model discussed previously and used to effectively optimize the system. Several investigations have been carried out along these lines ([24], [37], [18] to name just a few) and much more remains to be done. It is hoped that the present formulation can help in these future studies.

CHAPTER 7

Concluding Remarks

In this lecture, I have presented the algebraic theory of color spaces, and I have extended it to color image spaces. Although rooted in the work of Krantz, the approach is different and goes well beyond that work in many directions. I believe that the presentation here is unique, and I hope that it can form a good basis for future work in color imaging and color image processing.

Due to time and length constraints, this lecture does not contain many topics I would have liked to include and some of these items will have to remain as future work. Some of these directions include:

- The theory of transformations between different color spaces can be applied to more concrete scenarios, using it as a basis for the optimization of the transformation.

- The vector formulation of demosaicking can be used to develop new demosaicking algorithms, applying it to other CFA patterns, and perhaps using it to optimize the pattern. Indeed it could be combined with the theory of color space transformations to optimize CFA patterns with more than three color classes.

- The formulation for color mosaic displays can be used to select optimal display configurations as well as to develop better filters to convert from the stored image to the display driving signal.

- The topological and metric properties of the color space should be more completely developed and integrated with the theory presented here.

- Random field models for color images based on the vector representation introduced here can be developed.

I will continue to work on these problems, and perhaps some of my readers will as well.

APPENDIX A

Semigroups and Groups

This appendix provides definitions and notation for semigroups and groups, the basic algebraic structures with one operation.

1. A *binary operation* Δ in a set $\mathcal{A}$ is a function from $\mathcal{A} \times \mathcal{A}$ into $\mathcal{A}$. We write $x \Delta y$ for $x, y \in \mathcal{A}$.

   For example, addition $+$ is a binary operation on the set $\mathbb{Z}$ of integers.

2. A binary operation Δ on a set $\mathcal{A}$ is *associative* if $(x \Delta y) \Delta z = x \Delta (y \Delta z)$ for all $x, y, z \in \mathcal{A}$.

   For an associative operation, the parentheses are not required, and we can write simply $x \Delta y \Delta z$. Addition on $\mathbb{Z}$ is associative.

3. A binary operation Δ on a set $\mathcal{A}$ is *commutative* if $x \Delta y = y \Delta x$ for all $x, y \in \mathcal{A}$.

   Addition on $\mathbb{Z}$ is also commutative.

4. A *semigroup* is a set $\mathcal{A}$ with one associative binary operation Δ, and is denoted $(\mathcal{A}, \Delta)$. A semigroup is *commutative* if its binary operation is commutative.

   The set of strictly positive integers $\mathbb{Z}_+^*$ is a commutative semigroup.

5. Let Δ be a binary operation on a set $\mathcal{A}$. An element $e \in \mathcal{A}$ is a *neutral element* if $e \Delta x = x \Delta e = x$ for all $x \in \mathcal{A}$.

   The set of strictly positive integers $\mathbb{Z}_+^*$ under addition has no neutral element while the set of non-negative integers $\mathbb{Z}_+$ has the neutral element 0.

6. Let $(\mathcal{A}, \Delta)$ be a semigroup with a neutral element e. An element $x \in \mathcal{A}$ is *invertible* if there exists $y \in \mathcal{A}$ such that $x \Delta y = y \Delta x = e$. Such an element y is called the inverse of x, and it is unique if it exists.

7. A *group* is a semigroup $(\mathcal{A}, \Delta)$ with a neutral element such that every element of $\mathcal{A}$ is invertible.

   All these definitions can be found in Chapter I of [57], in particular in Sections 2, 4 and 7.

APPENDIX B

Equivalence Relations

1. Let $\mathcal{A}$ be a set. A relation R is a subset of $\mathcal{A} \times \mathcal{A}$. If $(x_1, x_2) \in R$, we write $x_1 \ R \ x_2$. Examples of relations on $\mathbb{R}$ are $x_1 = x_2$ and $x_1 < x_2$.

2. A relation is said to be an *equivalence relation* if it satisfies the following three properties:

reflexivity:	$x \ R \ x$
symmetry:	$x_1 \ R \ x_2 \Rightarrow x_2 \ R \ x_1$
transitivity:	$x_1 \ R \ x_2$ and $x_2 \ R \ x_3 \Rightarrow x_1 \ R \ x_3$

   Of the above examples, $x_1 = x_2$ is an equivalence relation and $x_1 < x_2$ is not.

3. Let $\sim$ be an equivalence relation on a set $\mathcal{A}$, and define $(z) = \{x \in \mathcal{A} \mid x \sim z\}$ for any $z \in \mathcal{A}$. This set is called the equivalence class containing z. Let (z) and (y) be two equivalence classes. Then either $(z) = (y)$ or $(z) \cap (y) = \emptyset$. A consequence of this is that if y is any element of (z), then $(y) = (z)$, i.e., any element of the equivalence class can be used as the class representative. A second consequence is that every element $x \in \mathcal{A}$ belongs to one and only one equivalence class. Thus, the set of equivalence classes forms a partition of $\mathcal{A}$.

   See, for example, Section 10, Chapter II of [57] for a complete treatment.

APPENDIX C

Vector Spaces

This appendix briefly summarizes some of the key results from linear algebra that are used in this lecture. A great more detail can be found in standard texts, such as [19]. I also recommend Chapter 0 of [59], from which I learned much of the material over 30 years ago in the handwritten edition.

1. A general vector space is defined with respect to a field K. In this work, we will only be concerned with real and complex vector spaces defined with respect to the fields $\mathbb{R}$ and $\mathbb{C}$ of real and complex numbers. The development is carried out for complex vector spaces, since it is then easy to restrict everything to be real. In this appendix, elements of a vector space are denoted in boldface with an arrow over them, e.g., $\vec{x}$. Several different types of vector spaces appear in the main text, but they don't use this notation.

2. A complex vector space $\mathcal{S}$ is a set on which two operations are defined,

   vector addition $+ : \mathcal{S} \times \mathcal{S} \to \mathcal{S}$,

   multiplication by a scalar $\cdot : \mathbb{C} \times \mathcal{S} \to \mathcal{S}$

   that satisfy the following properties:

   A. For each $\vec{x}, \vec{y}, \vec{z} \in \mathcal{S}$

(a)	$\vec{x} + \vec{y} = \vec{y} + \vec{x}$	commutativity
(b)	$\vec{x} + (\vec{y} + \vec{z}) = (\vec{x} + \vec{y}) + \vec{z}$	associativity
(c)	there is a unique $\vec{0}$ such that $\vec{x} + \vec{0} = \vec{x}, \quad \forall \vec{x}$	zero vector
(d)	for each $\vec{x} \quad \exists(-\vec{x})$ such that $\vec{x} + (-\vec{x}) = \vec{0}$	negative

   B. For each $\alpha, \beta \in \mathbb{C}, \vec{x}, \vec{y} \in \mathcal{S}$

   (a) $\alpha \cdot (\beta \cdot \vec{x}) = (\alpha\beta) \cdot \vec{x}$
   (b) $1 \cdot \vec{x} = \vec{x}, \quad \forall \vec{x}$
   (c) $\alpha \cdot (\vec{x} + \vec{y}) = \alpha \cdot \vec{x} + \alpha \cdot \vec{y}$
   (d) $(\alpha + \beta) \cdot \vec{x} = \alpha \cdot \vec{x} + \beta\vec{x}$

   Note that $(\mathcal{S}, +)$ is a group. The symbol "$\cdot$" for multiplication by a scalar can usually be omitted, writing $\alpha\vec{x}$ in place of $\alpha \cdot \vec{x}$.

3. Let $\vec{x}_1, \vec{x}_2, \ldots, \vec{x}_p \in \mathcal{S}$. An expression of the form

$$\alpha_1\vec{x}_1 + \alpha_2\vec{x}_2 + \cdots + \alpha_p\vec{x}_p = \sum_{i=1}^{p} \alpha_i\vec{x}_i$$

is called a linear combination. The set of all linear combinations of $\vec{x}_1, \vec{x}_2, \ldots, \vec{x}_p$ is denoted span$(\vec{x}_1, \vec{x}_2, \ldots, \vec{x}_p)$.

4. The vectors $\vec{x}_1, \ldots, \vec{x}_p$ are said to be linearly independent if and only if

$$\sum_{i=1}^{p} \alpha_i \vec{x}_i = \vec{0} \qquad \text{implies } \alpha_i = 0, i = 1, \ldots, p.$$

Otherwise, $\vec{x}_1, \ldots, \vec{x}_p$ are said to be linearly dependent. If $\vec{x}_1, \ldots, \vec{x}_p$ are linearly independent, then

$$\sum_{i=1}^{p} \alpha_i \vec{x}_i = \sum_{i=1}^{p} \beta_i \vec{x}_i \quad \text{implies } \alpha_i = \beta_i, i = 1, \ldots, p,$$

i.e., a vector can be expressed in only one way as a linear combination of p linearly independent vectors.

5. If there exist p linearly independent vectors $\vec{b}_1, \ldots, \vec{b}_p$ such that *any* element of $\mathcal{S}$ can be written as a (unique) linear combination of $\vec{b}_1, \ldots, \vec{b}_p$, then $\mathcal{S}$ is a finite-dimensional vector space of dimension p, and $\{\vec{b}_1, \ldots, \vec{b}_p\}$ is called a basis. Let $\mathcal{B} = \{\vec{b}_1, \ldots, \vec{b}_p\}$ be a basis and let $\vec{x} \in \mathcal{S}$ be expressed $\vec{x} = x_1 \vec{b}_1 + \cdots + x_p \vec{b}_p$. Then, we can conveniently represent $\vec{x}$ as a $p \times 1$ column matrix with respect to the basis $\mathcal{B}$ as

$$\mathbf{x}_{\mathcal{B}} = \begin{bmatrix} x_1 \\ \vdots \\ x_p \end{bmatrix}.$$

6. A *subspace* of $\mathcal{S}$ is a subset $\mathcal{R}$ of $\mathcal{S}$, which is itself a vector space, i.e., closed under addition and scalar multiplication. We denote this $\mathcal{R} \subset \mathcal{S}$. For any $\{\vec{x}_1, \ldots, \vec{x}_p\}$, span$(\vec{x}_1, \vec{x}_2, \ldots, \vec{x}_p)$ is a subspace of $\mathcal{S}$.

7. If $\mathcal{R}, \mathcal{T} \subset \mathcal{S}$, we define the subspaces

$$\mathcal{R} + \mathcal{T} = \{\vec{r} + \vec{t} \mid \vec{r} \in \mathcal{R}, \vec{t} \in \mathcal{T}\},$$
$$\mathcal{R} \cap \mathcal{T} = \{\vec{x} \mid \vec{x} \in \mathcal{R} \text{ and } \vec{x} \in \mathcal{T}\}$$

8. Let $\mathcal{R}_1, \ldots, \mathcal{R}_K$ be subspaces of $\mathcal{S}$. We say that $\mathcal{R}_1, \ldots, \mathcal{R}_K$ are independent if $\vec{r}_1 + \cdots + \vec{r}_K = \vec{0}$ for $\vec{r}_i \in \mathcal{R}_i$ implies $\vec{r}_i = \vec{0}$ for $i = 1, \ldots, K$. If $\mathcal{S} = \mathcal{R}_1 + \cdots + \mathcal{R}_K$, it follows that any $\vec{x} \in \mathcal{S}$ can be written in a *unique* fashion as

$$\vec{x} = \vec{r}_1 + \cdots + \vec{r}_K, \quad \vec{r}_i \in \mathcal{R}_i.$$

In this situation, we write

$$\mathcal{S} = \mathcal{R}_1 \oplus \mathcal{R}_2 \oplus \cdots \oplus \mathcal{R}_K$$

and say that $\mathcal{S}$ is the *direct sum* of $\mathcal{R}_1, \ldots, \mathcal{R}_K$.

9. Let $\mathcal{R}$ and $\mathcal{S}$ be vector spaces. A linear transformation from $\mathcal{R}$ to $\mathcal{S}$ is a function $\mathcal{T} : \mathcal{R} \to \mathcal{S}$ such that $\mathcal{T}(\alpha \vec{x} + \vec{y}) = \alpha \mathcal{T} \vec{x} + \mathcal{T} \vec{y}$ for all $\vec{x}, \vec{y} \in \mathcal{R}$ and all $\alpha \in \mathbb{R}$. Let $\mathcal{B} = \{\vec{b}_1, \ldots, \vec{b}_p\}$ be a basis for $\mathcal{R}$ and $\mathcal{D} = \{\vec{d}_1, \ldots, \vec{d}_q\}$ be a basis for $\mathcal{S}$. Then, a linear transformation is completely specified numerically by its action on the p basis vectors of $\mathcal{R}$, expressed with respect to the q basis vectors of $\mathcal{S}$. Let

$$ \mathcal{T}(\vec{b}_i) = \sum_{k=1}^{q} t_{ki} \vec{d}_k. \tag{C.1} $$

Then, if $\vec{y} = \mathcal{T}(\vec{x})$, we have

$$ \begin{aligned} \vec{y} &= \mathcal{T}\left(\sum_{i=1}^{p} x_i \vec{b}_i \right) \\ &= \sum_{i=1}^{p} x_i \sum_{k=1}^{q} t_{ki} \vec{d}_k \\ &= \sum_{k=1}^{q} \left(\sum_{i=1}^{p} t_{ki} x_i \right) \vec{d}_k \\ &= \sum_{k=1}^{q} y_k \vec{d}_k. \end{aligned} $$

Thus, $\mathbf{y}_{\mathcal{D}} = \mathbf{T}_{\mathcal{B} \to \mathcal{D}} \mathbf{x}_{\mathcal{B}}$, where $\mathbf{T}_{\mathcal{B} \to \mathcal{D}}$ is the $q \times p$ matrix $[t_{ki}]$ defined by Eq. (C.1).

The *kernel* (or null space) of $\mathcal{T}$ is the subspace

$$ \ker \mathcal{T} = \{ \vec{x} \in \mathcal{R} \mid \mathcal{T}\vec{x} = \vec{0} \} \subset \mathcal{R}. $$

10. Let $\mathcal{R} \subset \mathcal{S}$. We say that $\vec{x}, \vec{y} \in \mathcal{S}$ are equivalent mod $\mathcal{R}$ if $\vec{x} - \vec{y} \in \mathcal{R}$. This relation satisfies all the properties of an equivalence relation. The equivalence classes are called cosets of the subspace $\mathcal{R}$:

$$ \vec{x} + \mathcal{R} = \{ \vec{y} \mid \vec{y} - \vec{x} \in \mathcal{R} \} $$

As usual, two equivalence classes are either identical or disjoint and the set of all equivalence classes forms a partition of $\mathcal{S}$.

We define the *quotient space* (or factor space) $\mathcal{S}/\mathcal{R}$ as the set of all equivalence classes,

$$ \mathcal{S}/\mathcal{R} = \{ \vec{x} + \mathcal{R} \mid \vec{x} \in \mathcal{S} \}. $$

The set of equivalence classes forms a vector space under the operations

$$ \begin{aligned} (\vec{x} + \mathcal{R}) + (\vec{y} + \mathcal{R}) &= (\vec{x} + \vec{y}) + \mathcal{R}, \\ \alpha(\vec{x} + \mathcal{R}) &= (\alpha\vec{x}) + \mathcal{R}. \end{aligned} $$

It is straightforward and standard to show that these operations are well-defined and unambiguous. The function $\mathcal{Q} : \vec{x} \mapsto \vec{x} + \mathcal{R}$ from $\mathcal{S}$ to $\mathcal{S}/\mathcal{R}$ is a linear transformation called the quotient transformation or the canonical projection of $\mathcal{S}$ on $\mathcal{S}/\mathcal{R}$. As we see in the main text, a color space is a factor space of the space $\mathcal{A}$ of stimuli.

APPENDIX D

Lattices

Lattices have been widely used to describe sampled multidimensional signals with non-rectangular sampling structures. For the purposes of this lecture, we are concerned with discrete-domain two-dimensional still ($D = 2$) and time-varying ($D = 3$) images. This appendix summarizes the key concepts and notations used in the lecture. Detailed expositions and illustrations can be found in [6], [10] and [11]. Although the discussion is for general dimension D, we are mainly concerned with the cases $D = 2$ and $D = 3$.

1. A *lattice* Λ in D dimensions is the set of all linear combinations, with integer coefficients, of D linearly independent vectors $v_1, \ldots, v_D$ in $\mathbb{R}^D$,

$$\Lambda = \{n_1 v_1 + \cdots + n_D v_D \mid n_i \in \mathbb{Z}\}. \tag{D.1}$$

   The basis vectors v_i are expressed as $D \times 1$ column matrices, and thus so are the elements of Λ. Note that $(\Lambda, +)$ forms a group.

2. The $D \times D$ matrix $\mathbf{V} = [v_1 \cdots v_D]$ is referred to as a *sampling matrix* for Λ. Then, we write

$$\Lambda = \text{LAT}(\mathbf{V}) = \{\mathbf{V}n \mid n \in \mathbb{Z}^D\}. \tag{D.2}$$

   The sampling matrix for a given lattice is not unique; $\text{LAT}(\mathbf{V}_1) = \text{LAT}(\mathbf{V}_2)$ if and only if $\mathbf{E} = \mathbf{V}_1^{-1}\mathbf{V}_2$ is unimodular, i.e., an integer matrix such that $|\det \mathbf{E}| = 1$.

3. A unit cell of a lattice Λ is a set $\mathcal{P} \subset \mathbb{R}^D$ such that copies of $\mathcal{P}$ centered on each lattice point tile all of $\mathbb{R}^D$ without overlap. The unit cell is not unique. The area of any unit cell is $d(\Lambda) = |\det \mathbf{V}|$ for any sampling matrix $\mathbf{V}$. The *Voronoi* cell is a unit cell in which no point is closer to any non-zero element of Λ than to the origin $\mathbf{0}$.

4. The set $\Lambda^* = \{r \mid r \cdot x \in \mathbb{Z} \text{ for all } x \in \Lambda\}$ is a lattice known as the *reciprocal lattice*. If $\Lambda = \text{LAT}(\mathbf{V})$, then $\Lambda^* = \text{LAT}(\mathbf{V}^{-T})$ where $\mathbf{V}^{-T}$ denotes $(\mathbf{V}^T)^{-1}$, T denotes matrix transpose and $r \cdot x$ denotes the matrix product $\mathbf{r}^T\mathbf{x}$. $d(\Lambda^*) = 1/d(\Lambda)$.

5. Γ is a *sublattice* of Λ if both Λ and Γ are lattices, and every point of Γ belongs to Λ. We write $\Gamma \subset \Lambda$. $\Gamma = \text{LAT}(\mathbf{V}_\Gamma)$ is a sublattice of $\Lambda = \text{LAT}(\mathbf{V}_\Lambda)$ if and only if $\mathbf{M} = (\mathbf{V}_\Lambda)^{-1}\mathbf{V}_\Gamma$ is an integer matrix.

6. If $\Gamma \subset \Lambda$, then $\Lambda^* \subset \Gamma^*$.

7. If $\Gamma \subset \Lambda$, so that $\mathbf{V}_\Gamma = \mathbf{V}_\Lambda \mathbf{M}$ for an integer matrix $\mathbf{M}$, then $d(\Gamma) = |\det \mathbf{M}|d(\Lambda)$ where $K = |\det \mathbf{M}|$ is an integer. $K = d(\Gamma)/d(\Lambda)$ is called the index of Γ in Λ, denoted $(\Lambda : \Gamma)$.

8. If $\Gamma \subset \Lambda$, the set

$$c + \Gamma = \{c + x \mid x \in \Gamma\} \tag{D.3}$$

for any $c \in \Lambda$ is called a *coset* of Γ in Λ. Two cosets are either identical or disjoint: $c + \Gamma = d + \Gamma$ if $c - d \in \Gamma$; otherwise $(c + \Gamma) \bigcap (d + \Gamma) = \emptyset$. There are $K = (\Lambda : \Gamma)$ distinct cosets of Γ in Λ. If $b_1, \ldots, b_K$ are arbitrary representatives of these K cosets, we have

$$\Lambda = \bigcup_{k=1}^{K} (b_k + \Gamma). \tag{D.4}$$

9. Let $f(x), x \in \Lambda$ be a scalar signal defined on a lattice Λ. We define the *Fourier transform* of $f(x)$ to be

$$F(u) = \sum_{x \in \Lambda} f(x) \exp(-j 2\pi u \cdot x) \tag{D.5}$$

where u is a D-dimensional frequency vector, with components expressed in cycles per unit of length or time as appropriate. The Fourier transform is a periodic function of the continuous frequency vector, with periodicity given by the reciprocal lattice: $F(u) = F(u + r)$ for all $r \in \Lambda^*$. The Fourier transform of $f(x) \exp(j 2\pi u_0 \cdot x)$ is $F(u - u_0)$ for an arbitrary fixed frequency vector u_0; this is the modulation property.

10. A signal $f(x), x \in \Lambda$ is periodic with periodicity lattice Γ if $f(x + c) = f(x)$ for all $c \in \Gamma$, where $\Gamma \subset \Lambda$. There are $K = (\Lambda : \Gamma)$ distinct values of this signal, which form one period. These are $f(b_1), \ldots, f(b_K)$, where $\{b_1, \ldots, b_K\}$ is an arbitrary set of coset representatives of Γ in Λ. The periodic signal is constant on cosets of Γ in Λ.

11. A periodic signal $f(x), x \in \Lambda$, with periodicity lattice $\Gamma \subset \Lambda$ has the discrete Fourier series representation

$$f(x) = \sum_{k=1}^{K} F(k) \exp(j 2\pi x \cdot d_k), \quad x \in \Lambda \tag{D.6}$$

where

$$F(k) = \frac{1}{K} \sum_{i=1}^{K} f(b_i) \exp(-j 2\pi b_i \cdot d_k), \quad k = 1, \ldots, K. \tag{D.7}$$

In these expressions, $K = (\Lambda : \Gamma), b_1, \ldots, b_K$ are coset representatives for Γ in Λ and $d_1, \ldots, d_K$ are coset representatives for Λ^* in Γ^*.

Bibliography

[1] H. H. Barrett and K. J. Myers, *Foundations of Image Science.* Hoboken, NJ: Wiley-Interscience, 2004.

[2] H. Brettel, F. Viénot, and J. D. Mollon, "Computerized simulation of color appearance for dichromats," *J. Opt. Soc. Am. A, Opt. Image Sci.*, vol. 14, no. 10, pp. 2647–2655, Oct. 1997. DOI: 10.1364/JOSAA.14.002647

[3] J. B. Cohen, *Visual Color and Color Mixture: The Fundamental Color Space.* Urbana, IL: University of Illinois Press, 2001.

[4] J. B. Cohen and W. E. Kappauf, "Color mixture and fundamental metamers: theory, algebra, geometry, application," *American Journal of Psychology*, vol. 98, no. 2, pp. 171–259, 1985. DOI: 10.2307/1422442

[5] R. F. Dougherty and A. R. Wade. (2009) Vischeck. [Online]. Available: http://www.vischeck.com/

[6] E. Dubois, "The sampling and reconstruction of time-varying imagery with application in video systems," *Proc. IEEE*, vol. 73, no. 4, pp. 502–522, Apr. 1985. DOI: 10.1109/PROC.1985.13182

[7] E. Dubois, "Frequency-domain methods for demosaicking of Bayer-sampled color images," *IEEE Signal Process. Lett.*, vol. 12, pp. 847–850, 2005. DOI: 10.1109/LSP.2005.859503

[8] E. Dubois, "Filter design for adaptive frequency-domain Bayer demosaicking," in *Proc. IEEE Int. Conf. Image Processing*, Oct. 2006, pp. 2705–2708. DOI: 10.1109/ICIP.2006.313073

[9] E. Dubois, "Color filter array sampling of color images: Frequency-domain analysis and associated demosaicking algorithms," in *Single Sensor Imaging: Methods and Applications for Digital Cameras*, R. Lukac, Ed. Boca Raton, FL: CRC Press, 2009, ch. 7, pp. 183–212.

[10] E. Dubois, "Video sampling and interpolation," in *The Essential Guide to Video Processing*, A. Bovik, Ed. Academic Press, 2009, ch. 2.

[11] D. E. Dudgeon and R. M. Mersereau, *Multidimensional Digital Signal Processing.* Englewood Cliffs, NJ: Prentice-Hall, 1984.

[12] M. D. Fairchild, *Color Appearance Models.* Reading, MA: Addison-Wesley, 1998.

[13] J. Farrell, G. Ng, X. Ding, K. Larson, and B. Wandell, "A display simulation toolbox for image quality evaluation," *Journal of Display Technology*, vol. 4, no. 2, pp. 262–270, Jun. 2008. DOI: 10.1109/JDT.2007.913963

[14] D. Fearnley-Sander, "Hermann Grassmann and the creation of Linear Algebra," *The American Mathematical Monthly*, vol. 86, no. 10, pp. 809–817, Dec. 1979. DOI: 10.2307/2320145

[15] G. H. Golub and C. F. Van Loan, *Matrix Computations*, 3rd ed. Baltimore, MD: Johns Hopkins University Press, 1996.

[16] H. G. Grassmann, "Theory of compound colors," *Philosophic Magazine*, vol. 4, no. 7, pp. 254–264, 1854, translated from original German version which appeared in Annelen der Physik und Chemie (Poggendorf), vol. 89, pp. 69-84, 1853. Reproduced in D.L. MacAdam, Sources of Color Science, MIT Press, 1979, pp. 53-60.

[17] B. K. Gunturk, J. Glotzbach, Y. Altunbasak, R. W. Schafer, and R. M. Mersereau, "Demosaicking: Color filter array interpolation," *IEEE Signal Process. Mag.*, vol. 22, no. 1, pp. 44–54, Jan. 2005. DOI: 10.1109/MSP.2005.1407714

[18] K. Hirakawa and P. J. Wolfe, "Fourier domain display color filter array design," in *Proc. IEEE Int. Conf. Image Processing*, San Antonio, TX, Sep. 2007, pp. III–429–III–432. DOI: 10.1109/ICIP.2007.4379338

[19] K. Hoffman and R. Kunze, *Linear Algebra*, 2nd ed. Upper Saddle River, NJ: Prentice-Hall, 1971.

[20] J. Horváth, *Topological Vector Spaces and Distributions*. Reading, MA: Addison-Wesley, 1966, vol. 1.

[21] R. W. G. Hunt, *The Reproduction of Colour*, 6th ed. Chichester, UK: John Wiley & Sons, 2004.

[22] C. P. Huynh and A. Robles-Kelly, "Comparative colorimetric simulation and evaluation of digital cameras using spectroscopy data," in *Proc. 9th Biennial Conference of the Australian Pattern Recognition Society on Digital Image Computing Techniques and Applications*, 2007, pp. 309–316.

[23] G. M. Johnson and M. D. Fairchild, "A top down description of S-CIELAB and CIEDE2000," *Color Research and Application*, vol. 28, no. 6, pp. 425–435, Dec. 2003. DOI: 10.1002/col.10195

[24] M. A. Klompenhouwer and G. de Haan, "Subpixel image scaling for color matrix displays," *Journal of the Society for Information Display*, vol. 11, no. 1, pp. 99–108, Mar. 2003. DOI: 10.1889/1.1831726

[25] J. J. Koenderink and A. J. van Doorn, "Perspectives on colour space," in *Colour Perception: Mind and the Physical World*, R. Mausfeld and D. Heyer, Eds. Oxford, UK: Oxford University Press, 2003, ch. 1, pp. 1–56.

[26] D. H. Krantz, "Color measurement and color theory: I. Representation theorem for Grassman structures," *Journal of Mathematical Psychology*, vol. 12, pp. 283–303, 1975. DOI: 10.1016/0022-2496(75)90026-7

[27] R. G. Kuehni, "Color difference formulas: An unsatisfactory state of affairs," *Color Research and Application*, vol. 33, no. 4, pp. 324–326, Aug. 2008. DOI: 10.1002/col.20419

[28] R. G. Kuehni and A. Schwarz, *Color Ordered: A Survey of Color Order Systems from Antiquity to the Present*. New York, NY: Oxford University Press, 2008.

[29] G. Kutas, H.-K. Choh, Y. Kwak, P. Bodrogi, and L. Czúni, "Subpixel arrangements and color image rendering methods for multiprimary displays," *J. Electr. Imaging*, vol. 15, no. 2, pp. 023 002–1–023 002–9, Apr.-Jun. 2006. DOI: 10.1117/1.2193987

[30] H.-C. Lee, *Introduction to Color Imaging Science*. Cambridge, UK: Cambridge University Press, 2005. DOI: 10.2277/052184388X

[31] J. Lehtonen, J. Parkkinen, and T. Jaaskelainen, "Optimal sampling of color spectra," *J. Opt. Soc. Am. A, Opt. Image Sci.*, vol. 23, no. 12, pp. 2983–2988, Dec. 2006. DOI: 10.1364/JOSAA.23.002983

[32] D. L. MacAdam, *Sources of Color Science*. Cambridge, MA: MIT Press, 1970.

[33] A. W. Naylor and G. R. Sell, *Linear Operator Theory in Engineering and Science*. Ney York, NY: Springer, 1982.

[34] N. Ohta and G. Wyszecki, "Theoretical chromaticity-mismatch limits of metamers viewed under different illuminants," *J. Opt. Soc. Am.*, vol. 65, no. 3, pp. 327–333, Mar. 1975. DOI: 10.1364/JOSA.65.000327

[35] O. Packer and D. R. Williams, "Light, the retinal image, and photoreceptors," in *The Science of Color*, 2nd ed., S. K. Shevell, Ed. Amsterdam: Elsevier, 2003, ch. 2, pp. 41–102.

[36] A. Papoulis, *Systems and Transforms with Applications in Optics*. New York, NY: McGraw-Hill, 1968.

[37] J. C. Platt, "Optimal filtering for patterned displays," *IEEE Signal Process. Lett.*, vol. 7, no. 7, pp. 179–181, Jul. 2000. DOI: 10.1109/97.847362

[38] A. B. Poirson and B. A. Wandell, "The appearance of colored patterns: pattern-color separability," *J. Opt. Soc. Am. A, Opt. Image Sci.*, vol. 10, no. 12, pp. 2458–2470, Dec. 1993. DOI: 10.1364/JOSAA.10.002458

[39] A. B. Poirson and B. A. Wandell, "Pattern-color separable pathways predict sensitivity to simple colored patterns," *Vision Res.*, vol. 36, no. 4, pp. 515–526, 1996. DOI: 10.1016/0042-6989(96)89251-0

[40] C. Poynton, *Digital Video and HDTV Algorithms and Interfaces*. San Francisco, CA: Morgan Kaufmann, 2003.

[41] M. B. Priestley, *Spectral Analysis and Time Series: Univariate Series*. London: Academic Press, 1981, vol. 1.

[42] I. Richards and H. Youn, *Theory of Distributions: A Non-Technical Introduction*. Cambridge, UK: Cambridge University Press, 1990.

[43] J. Schanda, Ed., *Colorimetry: Understanding the CIE System*. Hoboken, NJ: John Wiley & Sons, 2007.

[44] F. J. M. Schmitt, "A method for the treatment of metamerism in colorimetry," *J. Opt. Soc. Am.*, vol. 66, no. 6, pp. 601–608, Jun. 1976. DOI: 10.1364/JOSA.66.000601

[45] E. Schrödinger, "Grundlinien einer Theorie der Farbenmetrik im Tagessehen. Teil 2," *Annalen der Physik*, vol. 63, no. 4, pp. 481–520, 1920, English translation in D.L. MacAdam, *Sources of Color Science*, MIT Press, 1970, pp. 155-182.

[46] G. Sharma, "Color fundamentals for digital imaging," in *Digital Color Imaging Handbook*, G. Sharma, Ed. Boca Raton, FL: CRC Press, 2003, ch. 1, pp. 1–114.

[47] G. Sharma and H. J. Trussell, "Decomposition of fluorescent illuminant spectra for accurate colorimetry," in *Proc. IEEE Int. Conf. Image Processing*, vol. 2, Austin, TX, Nov. 1994, pp. 1002–1006. DOI: 10.1109/ICIP.1994.413506

[48] G. Sharma and H. J. Trussell, "Figures of merit for color scanners," *IEEE Trans. Image Process.*, vol. 6, no. 7, pp. 990–1001, Jul. 1997. DOI: 10.1109/83.597274

[49] W. S. Stiles and J. M. Burch, "N.P.L. colour-matching investigation: Final report (1958)," *Journal of Modern Optics*, vol. 6, no. 1, pp. 1–26, 1959. DOI: 10.1080/713826267

[50] A. Stockman and L. T. Sharpe, "Spectral sensitivities of the middle- and long-wavelength sensitive cones derived from measurements in observers of known genotype," *Vision Research*, vol. 40, pp. 1711–1737, 2000. DOI: 10.1016/S0042-6989(00)00021-3

[51] A. Stockman and L. T. Sharpe. (2008) Colour and vision database. Color & Vision Research Laboratories, UCL Institute of Ophthamology. London, England. [Online]. Available: http://cvrl.ioo.ucl.ac.uk/

[52] H. J. Trussel and M. S. Kulkarni, "Sampling and processing of color signals," *IEEE Trans. Image Process.*, pp. 677–681, Apr. 1996. DOI: 10.1109/83.491346

[53] H. J. Trussell, "Application of set theoretic methods to color systems," *Color Research and Application*, vol. 16, no. 1, pp. 31–41, Feb. 1991. DOI: 10.1002/col.5080160108

[54] P. Urban and R.-R. Grigat, "Metamer density estimated color correction," *Signal, Image and Video Processing*, vol. 3, no. 2, pp. 171–182, Jun. 2009. DOI: 10.1007/s11760-008-0069-0

[55] P. Viénot, H. Brettel, L. Ott, A. Ben M'Barek, and J. D. Mollon, "What do colour-blind people see?" *Nature*, vol. 376, no. 6536, pp. 127–128, 13 July 1995.

[56] Y. Wang, J. Ostermann, and Y.-Q. Zhang, *Video Processing and Communications.*   Upper Saddle River, NJ: Prentice-Hall, 2002.

[57] S. Warner, *Modern Algebra.*   Englewood Cliffs, NJ: Prentice-Hall, 1965, vol. 1.

[58] K. Witt, "CIE color difference metrics," in *Colorimetry: Understanding the CIE System*, J. Schanda, Ed.   Hoboken, NJ: John Wiley & Sons, 2007, ch. 4, pp. 79–100.

[59] W. M. Wonham, *Linear Multivariable Control: A Geometric Approach*, 3rd ed.   New York, NY: Springer-Verlag, 1985.

[60] W. D. Wright, "Professor Wright's Paper from the Golden Jubilee Book: The historical and experimental background to the 1931 CIE system of colorimetry," in *Colorimetry: Understanding the CIE System*, J. Schanda, Ed.   Hoboken, NJ: John Wiley & Sons, 2007, ch. 2, pp. 9–23.

[61] G. Wyszecki and W. S. Stiles, *Color Science: Concepts and Methods, Quantitative Data and Formulae*, 2nd ed.   New York: John Wiley & Sons, 1982.

[62] J. Xu, J. Farrell, T. Matskewich, and B. Wandell, "Prediction of preferred ClearType filters using the S-CIELAB metric," in *Proc. IEEE Int. Conf. Image Processing*, San Diego, CA, Oct. 2008, pp. 361–364. DOI: 10.1109/ICIP.2008.4711766

[63] X. Zhang. (1998) S-CIELAB: A spatial extension to the CIE L*a*b* DeltaE color difference metric. [Online]. Available: http://white.stanford.edu/~brian/scielab/

[64] X. Zhang and B. A. Wandell, "A spatial extension of CIELAB for digital color image reproduction," *Journal of the Society for Information Display*, vol. 5, no. 1, pp. 61–67, 1997. DOI: 10.1889/1.1985127

[65] L. Zuppiroli, M.-N. Bussac, and C. Grimm, *Traité des couleurs.*   Lausanne, Switzerland: Presses polytechniques et universitaires romandes, 2003.

Author's Biography

ERIC DUBOIS

Eric Dubois received the B.Eng. (honours) degree with great distinction and the M.Eng. degree from McGill University in 1972 and 1974, and the Ph.D. from the University of Toronto in 1978, all in electrical engineering. He joined the Institut national de la recherche scientifique (University of Quebec) in 1977, where he held the position of professor in the INRS-Télécommunications centre in Montreal, Canada until 1998. Since July 1998, he has been Professor with the School of Information Technology and Engineering (SITE) at the University of Ottawa, Ottawa, Canada. He was Vice-Dean (Research) and Secretary of the Faculty of Engineering from 2001 to 2004. From January 2006 to December 2008 he was Director of SITE.

His research has centered on the compression and processing of still and moving images, and in multidimensional digital signal processing theory. His current research is focused on stereoscopic and multiview imaging, image sampling theory, image-based virtual environments and color signal processing. The research has been carried out in collaboration with such organizations as the Communications Research Centre, the National Research Council, the RCMP, and the Learning Objects Repositories Network (LORNET).

Dubois is corecipient of the 1988 Journal Award from Society of Motion Picture and Television Engineers. He is a Fellow of the IEEE, of the Canadian Academy of Engineering and of the Engineering Institute of Canada. He is a registered professional engineer in Quebec (member of the Order of Engineers of Quebec). He is a member of the Society for Information Display (SID) and the Society for Imaging Science and Technology (IS&T). He is a member of the editorial board of the EURASIP journal Signal Processing: Image Communication and was an associate editor of the IEEE Transactions on Image Processing (1994-1998). He was technical program co-chair for the IEEE 2000 International Conference on Image Processing (ICIP) and a member of the organizing committee for the IEEE 2004 International Conference on Acoustics, Speech and Signal Processing (ICASSP).

Printed in the United States
by Baker & Taylor Publisher Services